The APBC Book of Companion Animal Behaviour

The APBC Book of Companion Animal Behaviour

Edited by David Appleby

SOUVENIR PRESS

Published in Great Britain in 2004 by
Souvenir Press Ltd
43 Great Russell Street
London WC1B 3PD

Reprinted 2004

ISBN 0 285 63699 5

Typeset by FiSH Books, London
Printed in Great Britain by
Creative Print & Design Group (Wales), Ebbw Vale

Contents

List of Illustrations

11. Urine spraying – a natural method of communication for the cat, but unpleasant for humans if performed indoors

12. Some multi-cat households are made up of sibling pairs such as these, but in many cases there is no such basis to the feline community and cats are expected to live with total feline strangers

13. Rabbits are curious creatures and will play with and investigate objects in their environment

14. Garden and indoor toys can provide security, shade and stimulation

15. The rabbit's natural behaviour makes it possible to use a litter tray

16a and b. Creative thinking can provide environmental stimulation for dogs in a shelter environment

17. Environmental enrichment for the cat in a shelter should include different levels and places to hide

Introduction

The APBC's previous book *The Behaviour of Dogs and Cats*, (Stanley Paul, 1993) brought many important concepts to the fore and was a big step forward for its time. Over the last few years, the study of companion animal behaviour has advanced our knowledge and improved our ability to prevent and treat behaviour problems. These advances have occurred concurrently with increased exchanges of information internationally and cross pollination of ideas. Nonetheless, there is still a long search ahead in order to understand why behaviour problems develop and what treatments are most effective. This book reflects some of the current thinking of the issues, and is written by people who have to process this and other information during the course of treating behaviour problems.

The book focuses on the three main companion animals: the dog, cat and rabbit. This is because these are the species of mammal that, at present, are most commonly presented for the treatment of behaviour problems. Although the dog has been overtaken by the cat in its popularity as a pet, it is still the species most frequently seen in behaviour practice, and this is reflected in the number of chapters devoted to it in this book. To a large extent the principles involved – ethology, learning theory and degree of cognition, early experience and emotional states, welfare and law – can be related to many other species.

John Fisher, who edited the earlier APBC book, said in his introduction that at times 'my editor's pencil has trembled at the responsibility'. I now understand exactly how he felt. I would like to thank my colleagues within the APBC for contributing chapters on their own special interest, and to APBC administrator Pauline Appleby for deciding the time was right to produce another book and driving it forward to publication.

David Appleby, Editor
2004

1
Evolutionary Aspects of Canine Behaviour

Jan Hoole

Canine Ecology

Dogs are carnivorous, predatory and social animals, and, like other canine species, are ecologically near the top of the food web. Wolves, from whom dogs are descended, are very successful predators, and as with all organisms, the ecological niche they occupy affects their morphology (physical appearance) and behavioural and social structure. As a direct descendant of the Grey Wolf (*Canis lupus*), domestic dog (*Canis familiaris)* shows a physiology and behaviour that is strongly influenced by it. Although humans have radically altered dog morphology through selective breeding, the physiology and some aspects of instinctive behaviour remain relatively unaltered.

To make sense of the evolution of the dog, we must first understand how the wolf behaves and how it interacts with its environment (its ecology). How natural selection affects behaviour is also important. We shall look at certain aspects of the wolf's behaviour and how these behaviours have been translated into our domestic companions.

Wolf Ecology, Morphology and Social Systems

The most striking characteristic of the wolf is its adaptation to running. This is true of most of the dog family, much more so than any of the other large carnivores. Of all the Canidae, wolves have the longest legs and the greatest endurance when running down prey. They tend to hunt prey that is much larger than they are, and so require help with the kill. Although it does happen from time to time, an individual wolf finds it difficult to bring down an adult reindeer or musk ox, and impossible to tackle moose or bison. By hunting co-operatively they can isolate a weak animal from the herd and join together to bring it down and kill it.

Wolves themselves are not small animals; they are between 26 and 32 inches (66 and 81cm) high at the shoulder, and males can be as much as 6 foot 5 inches (1.95m) long from nose to tail tip. There is, however, great variation, not only between species of wolves, but also within species, in size, shape, coat and behaviour. Since wolves are exceedingly difficult to observe in the wild little is known about the behaviour of many species, but the Grey Wolf has been the subject of many studies and its social and predatory behaviour has been reasonably well documented. Due to the secretive nature of the animal much of the work has been done on captive and semi-wild populations. However, even aspects of behaviour that have been seen many times are often the subject of conjecture about how they should be interpreted.

Status (Uninhibited/Inhibited Social Behaviour)

The wolf pack has a hierarchical social structure. Each individual has a particular status in relation to every other individual, and the structure tends to remain reasonably stable over quite long periods. The male and female wolves with the highest status tend to form a pair and to lead the pack. However, the common perception that they are a breeding pair, and are the parents of all the other pack members is not always true. There have been several recorded instances in which the second ranked male (usually called the beta male) was actually the sire of all the pups produced, while the alpha male did not mate successfully at all. In at least one of these situations, when the alpha male was removed and the beta male became the leader, he ceased to mate with the alpha female and the new beta male became the breeding male.

Status within packs tends to be separated into male and female lines. Within these lines there is a fairly simple 'pecking order' that is established as individuals reach social maturity. Pups and immature animals have a separate status structure that is not determined by gender. Some adults who are very low in status may not even be part of the social structure of the pack and remain on the outskirts. Dogs living in packs, such as hounds, huskies and feral dogs, and those living in a 'mixed pack' with humans also have status hierarchies, the maintenance of which is often complicated by human ignorance of canine etiquette and communication.

Communication

Being large and well armed animals, if wolves fought over resources within the pack they could do each other serious injuries, and this would not contribute to the overall health of the pack as a reproductive unit. Encounters within the pack are often governed by ritualised signals and symbolic postures that help the pack members to interpret each other's status and state of mind and thus avoid genuine aggression. In order to maintain harmony and stability within the pack, wolves are very good at reading each other's body language and posture, and they appear to use their own bodies to signal their state and intentions to other wolves. Whether or not such signalling is conscious or intentional is a matter for much discussion among academics. However, even if the signals are simply an unconscious response to the animal's internal state, they may enable other individuals to assess the situation and make a judgement about the likely behaviour of the animal.

Dogs maintain an excellent ability to read the signals emitted by humans and other dogs; however, selective breeding has altered many of the structures used for signalling. Ear shape is frequently changed from pricked to long and floppy, with which it is difficult to signal inhibited behaviour. Paedomorphosis (having the appearance of a new-born pup) is a popular characteristic in modern breeds, such as Cavalier King Charles Spaniel, Shih-tzu, Boxer and Chihuahua, and causes the dog to have a short muzzle, very round, prominent eyes and, often, dropped ears. Some breeds have very curled tails, or have their tails docked at birth, compounding the difficulties of signalling to other dogs.

A study carried out in 1997 found that the degree of paedomorphosis in individual breeds indicated the wolf developmental stage from which the signalling repertoire displayed by the breed was taken (Goodwin *et al.*, 1997). Those breeds that most closely resemble adult wolves have the greatest number of wolf signals, some of which are attained by wolves in the later stages of development, while breeds that look more like young pups have fewer and more infantile signals.

Many centuries of domestication, of course, separate dogs and wolves, and although they are sufficiently similar genetically to interbreed and produce fertile offspring, they are in fact, different species.

What is Evolution?

The process of evolution operates on all living organisms over long periods of time, usually many thousands or millions of years. In the wild the process is thought to be mediated by natural selection, but in the case of domestic animals the selection is artificial, the pressure applied by human intention rather than by natural forces such as changes in the environment. By this route all the domesticated animals known to humankind have emerged, often from an ancestral species that is now extinct. In the case of the dog, its ancestor, the Grey Wolf, is still around, just hanging on in the wilder areas of the earth.

Natural selection is a phrase that is often used to describe the means by which species arise. As a process it operates on all aspects of an organism: physical, ecological and behavioural. To understand how certain behavioural traits of the dog arose, we must consider the nature of its ancestor the wolf, and the possible evolutionary route by which the wolf acquired certain behaviours still observable in the dog. Evolution, whether by natural or artificial selection, will be more easily explained in the light of a basic understanding of how traits are inherited.

Inheritance

The basic unit of inheritance is the gene. Every cell that has a nucleus in a living organism contains all the genetic information needed to replicate the whole organism. Each time a cell divides, the genetic information is passed to the new cell by means of deoxyribonucleic acids (DNA), which is the form in which genetic information is stored. When a cell replicates itself the DNA is copied into ribonucleic acid (RNA), which is then translated into amino acids, the building blocks of proteins, and these are strung together to form proteins. The protein chains are biochemically joined to form biologically active molecules, membranes and cells.

Sometimes, in the process of copying or translation or of building the protein chain, an error occurs. Fortunately the cell has efficient mechanisms for spotting and repairing such mistakes, and most of them are repaired straight away. Occasionally an error is missed, and this is called a mutation. The effect on the organism will vary depending on where on the gene such a mutation is situated. In some areas of the DNA a mutation may not affect the organism at all. If the mutation is in a somatic (body) cell, it may have the effect of causing the cell not to die at the proper time, or to continue to replicate unnecessarily. This can result in a tumour. However, if the mutation happens to be in a germ cell, the sperm or egg by which an organism reproduces, the effect will be felt by the next generation rather than by the existing organism.

A mutation in a germ cell may result in a fertilised egg that is not viable, or cause the juvenile animal to have a genetic defect, which will mean that it is not able to survive to reproductive age or is sterile. In these cases the defect will not be passed on. The mutation may, however, be advantageous and give the organism an advantage in survival or reproductive strategy over other individuals in the population. These are the mutations that tend to be inherited and to become part of the gene pool of an organism.

From this extremely brief, and necessarily simplistic explanation it can clearly be seen that genes code for proteins, *and nothing else!* How, then, can they affect the behaviour of an organism? It is easy to see in real life that a Border Collie from a particular blood-line can have an inherited tendency to herd living things, which develops to a certain level without training by the owner. But it is more difficult to understand how the DNA in that collie's cells, which only translates into the building blocks of molecules, can cause the dog to wish to gather other animals into a herd.

Genotype and Phenotype

The particular version of its DNA that an organism carries in its cells is known as its *genotype*. When this is translated into the physical being that is a plant or animal it is called the *phenotype*. The phenotype is made of millions of cells, each of which relies on biochemical processes to keep it alive and enable it to perform its particular function. Many different biochemically active molecules, such as enzymes, hormones and neurotransmitters are involved in the regulation of these processes, and for each type of molecule there must be receptors within or on the surface of each cell in order to allow the cell to react to its presence. The number, position and sensitivity of these receptors is controlled by genes within the cell, and these genes may be 'switched on' or off at different times in the life of the cell in response to external or internal triggers. This means that changes within the body or in the external environment can cause the animal to respond in different ways to the chemicals that are present in its body, and that this response can be affected by inherited information.

Examples of how this might work can be seen in many species of birds and animals. If an animal has a greater number of receptors for a neurotransmitter that is calming to its nervous system, or the ability to produce more of that neurotransmitter than others of its species, it may be bolder, and more willing to explore new territory or new environments. An organism that has fewer receptors for that chemical or more for a neurotransmitter that is excitatory may be more anxious and tend to avoid novel stimuli. It is not likely to explore a fresh ecological niche, and its ability to form early pair bonds may be less than others of its species, so that its reproductive success may be compromised. There have been suggestions that black-coated rabbits are bolder and willing to remain above ground to feed longer than agouti-coated ones, enabling them to take in more nourishment. Melanistic (black) phase Great Skuas (a very large sea bird) may form pair bonds more easily than lighter coloured Skuas, therefore they begin nesting earlier, are able to choose the best nest sites and lay their eggs before other pairs, giving them a reproductive advantage.

The passage of genetic information is essentially a one-way process. The sequence of the DNA in the nucleus of a cell can affect the size and shape of the molecules produced, but the size and shape of the molecule does not influence the sequence of the DNA stored in the nucleus. It is not possible for a learned behaviour to be inherited, because the changes brought about in the nervous system by learning cannot be fed back into the genes. However, there is the possibility that an organism may be genetically predisposed to have the ability to learn a particular behaviour or type of behaviour, and this can be inherited. If the environment in which the animal finds itself offers the opportunity to learn a behaviour that ultimately contributes to the animal's reproductive success, the animal is more likely to be able to learn it. This will allow it to have more offspring, some of which will carry the gene for being able to learn the same thing.

It is a general genetic predisposition to behave in a certain way that is thus passed on and over generations is selected for in such a way that it becomes more narrowly adapted to the environmental conditions or ecological niche of an organism. Eventually the animal may become so behaviourally and morphologically adapted to its niche that it is incapable of behaving in a different way or of surviving in a different environment. It may also lose the ability to mate with organisms that are not so adapted, and thus a new species is created.

Selection Pressure and Variation

So natural selection can only operate if there is variation within a species. If all the organisms within a species are identical, as in clones, all will respond in the same way to changes in their environment and none will have an advantage. The genome of every organism contains an enormous quantity of DNA that is never used. This may consist of genes that are recessive, and are not expressed if a dominant allele (copy of a gene) is present. Sometimes they are genes that are 'switched off' and will only be expressed under certain circumstances or environmental conditions. In other cases they are genes that are redundant because the particular protein for which they code is no longer required in the phenotype of the organism. In most organisms these genes are present in multiple copies, only a few of which are expressed, and this is an excellent opportunity for mutations to take place that are not lethal to the organism. Together with mutations that occur spontaneously or in response to mutagens (substances, such as ultra-violet light or certain toxins, that cause mutations), the unexpressed genes in the genome are a source of variation.

Some of the small variations between individuals may not be apparent when the environment is favourable, but may cause an individual to behave in a slightly different way under differing conditions. For example, the size of nest that mice build is genetically determined, and is influenced by temperature. At 5°C the difference in nest size is much greater than when they are kept at 26°C. It is easy to imagine that, in a year in which the summer weather was poor, the mice that built larger nests which would keep their offspring warmer would have greater reproductive success than mice building smaller and less adequate nests. This would result in more of the large-nest-building gene being present in the next generation of the population.

Arctic Wolves generally sleep a short distance apart from each other at night, tightly curled into a ball with their noses covered by their tails. It might be expected that they would huddle together for warmth in the bitter arctic night, but L. David Mech points out that if they did this their breath would cause each other's coats to become damp and to freeze to the ground. It is easy to imagine that individuals in which preferred to sleep close together would more often die from this cause before reaching maturity, therefore failing to pass on their genes. Those which had a tendency to be wary of their fellows and preferred to sleep a little way apart would survive and pass on the inclination for aloof behaviour to their offspring.

Kin Selection

In a wolf pack, for most of the time only one pair produces pups. Occasionally a lower-ranking female will mate and succeed in bearing her pups, but generally a pack consists of two or three older animals and their offspring from successive years. The individuals which have not mated nevertheless take a share in the care of the current litter, often bringing food for them, helping to guard them and even to move them when the dam decides to change dens. This altruistic behaviour may seem difficult to understand. Surely they should be putting their efforts into attempting to pass on their own genes?

In fact, a mammal carries an average of 50 per cent of the same genes as its parents or full siblings and 25 per cent of an aunt, uncle or grandparent. Another way of putting this is to say that the probability of an individual sharing a particular gene with its parents, offspring or siblings is 0.5, and with its aunts, uncles and grandparents is 0.25. This measure is called the 'coefficient of relatedness'. Under some circumstances it is more advantageous to an individual's fitness, in terms of its chances of passing on its genes, for the animal to help its close relatives to survive and reproduce than to try itself to reproduce when there is small chance of success. This altruistic behaviour of helping relatives to survive or reproduce is called 'kin selection', and is the most likely evolutionary reason why wolves live in packs.

A lone pair of wolves has less chance of rearing pups to maturity in a harsh environment because it is more difficult for them to hunt large, energy-rich prey. While the female is in whelp and is nursing the pups the male must hunt alone and, although a single wolf is capable of killing an adult deer, caribou or even an elk, the risk involved in such encounters is greater for a lone wolf than for a pack. If the male is killed or injured the female would eventually have to abandon the pups in order to hunt, leaving them vulnerable to predators. However, if the male only hunts small prey, such as arctic hare or beaver, the risks are reduced, but the energy expended in hunting per prey item is likely to be much greater. By remaining with the parents and helping them to raise other pups, an individual wolf increases the chance of some of its genes surviving in the population.

One of the prerequisites for kin selection is that an animal is able to recognise its relatives, and there is evidence that some species, such as mice, are able to do this by recognising differences in a protein that occurs on all cells, the major histo-compatibility complex (MHC). The MHC enables the immune system to recognise cells that are part of the 'self' and therefore are not eliminated. A classical example of what can happen when MHCs are not compatible is rejection of transplanted tissues. It is believed that animals are able to recognise slight differences in scent which may correspond to differences in the MHC. It is possible that they can also detect similarities within a kinship line, and thus to whom they are related.

Wolves may also use visual cues to help recognise each other. Anyone who has kept more than one individual of the same breed of dog will know that, although they may look the same to outsiders, to the eye of familiarity it is scarcely possible to mix them up. Indeed it is possible to distinguish individuals by small details such as the pitch of a bark or the sound of a particular dog taking a drink.

Domestication

One of the more intriguing questions about the wolf concerns how this large and potentially dangerous predator became engaged with humans to such an extent that it eventually turned from *Canis lupus*, the wolf, into *Canis familiaris*, the domestic dog. In addition to considerable argument as to how this happened, biologists have expended a great deal of energy discussing when the event occurred.

The earliest identifiable dog remains in the fossil record are only 14,000 years old, while the oldest complete skeleton that can be said to be that of a dog was found in the grave of a woman in Israel, dated at 12,000 years ago. The use of molecular technology has enabled scientists to examine many theories about the origins of the dog, and a surprisingly large range of possibilities has been the result. Some studies have broadly agreed with the fossil record, putting the date of the earliest domestication event at around 15,000 years ago. Less conservative estimates suggest that it may have been as long ago as 135,000 years ago.

The consensus as to the genetic origins of the dog seems to be that the Grey Wolf *Canis lupus* is the common ancestor of all domestic dogs. Even the most ancient breeds found in the New World, such as the Xoloitzcuintli or Mexican hairless dog, came from stock imported from Eurasia.

Perhaps it is of even more interest to speculate on the 'how' of domestication. It is easy to imagine that ancient hunters, having killed an adult wolf, could have taken the pups home to rear by hand and thus provided the first nurturing interaction between man and wolf, but less easy to imagine why they might do such a thing.

While we may view wolves with sympathy in the present time, to a hunter in ancient times they would simply have been dangerous rivals, subsisting on similar prey to that utilised by humans. Unless one had already had experience of the process of socialisation of a wolf pup to humans and the profound effects it can have on the relationship between the two, to guess that socialisation is possible would be a leap of the imagination that would be beyond most humans. There are far more pressing reasons not to introduce a large predator into the family home, however appealing it may be at the puppy stage.

However, the effects of socialisation on wolf pups have been well documented, and it has been shown that hand-rearing can produce wolves that are not only tame but very friendly towards people. Once such wolves were established in a human settlement and began to breed, the pups would naturally be in contact with people from birth and so would come to view them as members of the pack. Over generations, the wolves most likely to survive would be the ones that remained tame, cooperated with the humans and had a high degree of inhibited behaviour. Individuals that showed aggression would be killed. This sort of selection process can profoundly alter unrelated parts of the genome. Over millennia, the dog has become an animal that comes into oestrus (becomes receptive to mating) twice as often as a wolf, has a smaller brain to body mass ratio, less powerful jaws, compacted teeth, and is more juvenile in its behaviour than its lupine ancestor.

An alternative theory is that when humans began to live in camps and settlements, opportunist wolves may have scavenged the refuse that would accumulate in the area. These would naturally be animals that were sufficiently bold to come into the proximity of a camp, and in a hard season, it is possible that the extra food source would help them to breed successfully, possibly with other scavenging wolves. The offspring of such wolves would undoubtedly learn from their parents about the supplementary food to be found in the vicinity of humans, and habituation from an early age would probably encourage them to go ever closer to the camps. Over a number of generations such wolves would eventually begin to show stronger traits of boldness and tameness, until it became possible for humans to approach them and form relationships with them.

This theory may be borne out by a long-running experiment carried out in Russia on silver foxes. These were bred for forty generations, the only criterion for allowing them to breed being tameness. After about twenty generations a female appeared which was very dog-like in temperament. She seemed to seek human company actively, and was quite playful. These traits bred true in her offspring and produced a line of foxes that resembled dogs in their tameness. It seems that this was due to a mutation that arose in that particular fox which produced a sudden change in levels of tameness, rather than to a gradual increase in tameness over generations. It is interesting that, although appearance was never bred for in the foxes, several dog-like characteristics, such as a white tip to the tail, seemed to be present in the line of tame individuals that had not been seen in the original stock. Clearly not one gene but several areas of the genome have been affected through breeding for one character, and this illustrates how closely linked can be genes that code for entirely different aspects of the organism.

Another circumstance that seems to support this theory has been documented in a city in Transylvania. A female wolf moved into woodland overlooking the city and began to raise a litter. She was radio-collared by some scientists who wished to monitor her feeding habits. They found that she not only visited the local shepherds with their flocks every night in the hope of finding a stray sheep, but had also taken to making regular trips to the municipal dump to scavenge refuse. For a lone wolf, this is an easier option than attempting to catch large prey that could cause injury, or small prey that is very energy-intensive to catch and gives low returns.

Behavioural similarities between dog and wolf

By whatever route domestication occurred, it is undeniable that many of the behavioural traits that we see in the dog can also be observed in the wolf. One of these is the greeting ritual. Most dogs, when their owners return home, even after a short absence, greet them with great enthusiasm, wagging the tail, wriggling and often attempting to jump up to lick the owner's face. In wolves a similar ritual is seen when the leader rejoins the pack after an absence. The other wolves will gather round the high-status individual, licking and nipping his or her muzzle, wagging their tails and displaying submissive body language. An individual wolf will also greet a higher-status individual in this manner, and the pack sometimes engages in a group 'greeting ritual' before a hunt.

The origins of this behaviour seem to lie in the food begging that pups engage in between weaning and the age when they are old enough to feed from the carcass. For this period adult wolves bring food to the pups, carrying it in the stomach. The pups stimulate the adults to regurgitate the food for them by frenzied licking of the adult's mouth. It seems that this behaviour is carried on into adulthood as a ritualised greeting to a higher-status (less inhibited) individual. Another purpose of greeting behaviour in canines in general is to ensure that nothing has changed with regard to status since the individuals separated. There is always the possibility that one of the wolves in the group has been injured or fallen ill since leaving the last greeting ceremony, and that may result in a general reappraisal of status within the pack.

When a dog wishes to engage in a greeting ritual with human members of its 'pack' it has the major disadvantage that its face is between 3 and 6 feet lower than that of its owner, depending on the breed of dog and the height of human. The dog naturally tries to get as close as possible to the mouth of the person and jumps up to reach it. After initial reinforcement during puppyhood, this is not usually well received by the humans, and can become a problem behaviour.

Vocal communication can be quite complex in both dogs and wolves. Although vocal communication is often divided into barking, growling, whining and howling, this does not adequately indicate the range of sounds produced and the information apparently conveyed by them. One of the most recognisable sounds produced by wolves is the howl. In Canada at certain seasons hundreds of people drive many miles into the wilderness to hear wolves howling, and recordings of 'wolf song' have been popular for many years in the music industry.

Wolves howl in several different circumstances. First when the pack is separated an individual will often howl and this attracts other members to it. Packs may also howl before a hunt, and this seems to be a matter of increasing the social cohesion of the pack at a time when the need for cooperation may be at odds with the individual's instinct to hunt for itself. In captive wolves the amount of howling is correlated with the degree of isolation of the wolf from people and from other wolves. It has been suggested that howling in these circumstances is an expression of loneliness, which may be true, but it is also likely that the individual is attempting to reunite the pack by calling others to it.

It is no coincidence that domestic dogs howl most frequently when left alone for long periods. Separation problems are dealt with elsewhere, but this attempt to call the pack home is directly traceable to the behaviour of wild wolves reuniting scattered pack members after hunting. Although we may suppress the behaviour in our companion animals by training them to cope with being alone, the unease that is generated in social animals by isolation is still present in dogs.

To return to the Border Collie, mentioned earlier, the herding behaviour it displays can also be traced to its origins in the hunting behaviour of the wolf. The hunting sequence in a predator can be separated into several distinct parts. Beginning with the animal displaying a fixed gaze towards the prey, known as 'eye', moving on to stalk, chase, pounce and kill. In some breeds of dog individual components of the sequence have been exaggerated by selective breeding until they are displayed to the exclusion of other parts. In the collie, 'eye' and 'stalk' are the most intensely displayed components, with the result that sheep faced with an apparent predator tend to bunch together and move away. The collie will often run round a herd of sheep or flock of geese, and this behaviour may have its origins in the predatory intention to find the weak animal in the herd. Certainly, the greatly prized ability to 'cut out' marked sheep in the shedding ring at sheepdog trials can be seen as a reflection of the wolf behaviour of singling out the vulnerable prey animal. However, although working sheepdogs are trained to, and may be genetically predisposed to, inhibit the latter parts of the predatory sequence, especially 'pounce' and 'kill', it should not be forgotten that it remains part of the behavioural repertoire of the breed, as of many other breeds.

Some claim, that the dog may have started to diverge from its ancestor, pre domestication, as far back as 135,000 years ago (Vila *et al.*, 1997), but even that is a very short time in evolutionary terms. Most of the changes in morphology of the domestic dog to create the very distinct breeds we now see were brought about within the last 150 years. The shape of the animal may have changed drastically in some cases, and certain aspects of the basic behaviour, such as the stalking instinct in collies, have been deliberately enhanced. The dog may have a small brain and little independence compared to its wild ancestors, but the underlying behavioural structure remains that of the wolf. Successful coexistence with a dog depends to some extent on the willingness of the human half of the partnership to become familiar with and to abide by the social structures and methods of communication of the wolf-pack.

2
The Foundations of Canine Behaviour

David Appleby and Jolanda Pluijmakers

Although the modern dog successfully fulfils the role of companion, society places pressure upon it to meet its expectations about its behaviour and yet that same society frequently leaves dogs ill equipped to achieve these ideals. As a consequence many dogs are unable to cope with everyday situations without resorting to avoidance or defensive behaviour. The welfare of these dogs is at risk. They cannot relax and enjoy life; they feel threatened by everyday events, and are susceptible to stress and disease. They are less likely, as a result, to make rewarding pets and are at risk of being abandoned, re-homed or euthanased.

A review of cases by the Association of Pet Behaviour Counsellors showed that, aggression motivated by fear towards people (25 per cent) and dogs (15 per cent) forms the most common reasons why dogs were referred for treatment by their vets, fears and phobias, such as in response to sounds, were reasons why a further 8 per cent of the cases analysed were referred. These figures probably represent the tip of an iceberg because many dogs which display problem behaviour related to fear do not have owners who seek help from a behaviour counsellor.

A dog's temperament and character are dependent upon genetic and environmental factors. The extent of the influence of each is not clear: the old 'nature-nurture' argument. This is not the impasse that is often implied because it is safe to say that genetics govern the potentials the dog can achieve, and experience and resultant learning shape behaviour and determine whether those potentials are realised. This shaping occurs throughout the dog's lifetime but there is a biologically pre-programmed period in early life during which simple exposure to things, or experience of interacting with them, has greater influence on the development of emotional states and behaviour than in later life. During this time, which is known as 'the sensitive period', the puppy learns what to expect and how to react. The results of this are a sense of control and a capacity to remain in a positive emotional state in varying situations. This organisation of a puppy's behaviour forms the foundation on which the development of its subsequent behaviour is based.

Current understanding of the sensitive period stems from research conducted in the 1950s, 1960's and 1970's. This suggested that fear of unfamiliar things or reduced willingness to approach them develops in identifiable phases between three to twelve–fourteen weeks of age. However, the interpretation of the results may have been misleading because the researchers depended upon observing changes in behaviour in response to the presence of unfamiliar stimuli. These changes occur after the relevant brain structures have developed and integrated, and following a period of learning that results in the organisation of the puppy's behaviour. Therefore the sensitive period may be much shorter than the researchers assumed. So when looking at the sensitive period from the perspective of emotional development the most important phase is between three to five weeks of age.

Emotional States

It is generally accepted that animals have emotional states such as fear, euphoria, anxiety etc. Whether they experience them consciously remains open to question, although the balance of evidence suggests that they do. Emotions enhance an individual's chances of survival by improving its ability to recognise events that are useful or unsafe, and to make appropriate responses. Situations that create a mismatch, or conflict with an individual's interests, give rise to negative emotions, such as fear, that require them to act in order to cope.

Whether an emotion is experienced as negative or positive is dependent on the activity of the dog's autonomic nervous system, which is controlled by the brain. This system is functionally divided into two parts: arousal (sympathetic) and relaxation (parasympathetic). These have mutually exclusive actions. Conditions of stress trigger activity in the sympathetic autonomic nervous system that has an emergency function, preparing the body for action. The effect includes increased heart rate, increasing blood supply to the muscles, heart and lungs, increased respiration and bowel movement, sometimes resulting in loss of toilet control, and chemical reactions such as the release of hormones. This activity co-occurs with negative emotional states, described in humans as fear, anxiety etc. A positive emotional state (parasympathetic) has a recuperative function, due to its opposite effect e.g., by restoring the blood supply and heart rate to normal.

Emotional Homeostasis

Every individual strives to achieve and maintain an emotional homeostasis, i.e., a positive emotional state. The function of emotional homeostasis is to allow an individual to deal with and adjust to the many changes that are part of everyday life. If every change in the environment caused a loss of emotional homeostasis the individual would never be able to relax and its welfare would be seriously impaired. To prevent this, a 'picture' of the world is formed early in life which sets the standard for situations during which they can afford to remain relaxed. Changes in the environment that deviate from this standard may cause a negative emotional state.

Maintenance Set

A dog's ability to remain in emotional homeostasis develops throughout the sensitive period for behavioural organisation; during part of this process mental representations of stimuli are formed and associated with responses. Those associated with a positive emotional state become part of a set of animate and inanimate objects that enable it to maintain emotional homeostasis and organise its behaviour. The effect on subsequent behaviour was shown in three-week-old Chihuahua puppies fostered individually in litters of four-week-old kittens (Fox, 1971). By twelve weeks of age, the cat-reared puppies preferred contact with cats over contact with other puppies which had not been fostered. Another experiment involved litters of puppies split into three groups. One group was hand-reared from three days old and received no canine contact. The second group was given equal canine and human contact. The third group only experienced other puppies and their dam. When these three groups of puppies were brought together at twelve weeks they showed a preference for puppies which had received similar rearing experience, and the puppies raised in isolation from other dogs had a poor relationship with other puppies (Fox & Stelzner, 1967).

Some maintenance stimuli grow to be so important that the dog becomes dependent on their presence to maintain emotional homeostasis. This dependency can be formed to any social or non-social stimulus, as is illustrated by these experiments. Elliot and Scott (1961) found that puppies confined in a strange pen were three times more distressed than puppies confined in a familiar pen. Cairns and Werboff (1972) housed four-week-old puppies with adult rabbits and found that within twenty-four hours of cohabitation, the puppies lost emotional homeostasis when separated from their rabbit, which was expressed by intense distress vocalisations, and attempts to escape and regain contact with the removed rabbit.

Dependency on social stimuli is more likely to develop than other stimuli, because they provide more stimulation and easily become associated with important factors such as contact comfort, suckling or play. The duration of exposure to a stimulus and its availability are important factors in determining dependency. If a maintenance stimulus, e.g., a female owner, is present in a variety of contexts, there is an increased likelihood that the dog will become dependent on her compared to other family members who are only present in some situations. A maintenance set that provides a dog with the best chance of maintaining an emotional homeostasis consists of a large number of stimuli some of which are of comparable worth. This is necessary because the composition of the set alters with changes in the environment and situation the dog is in. The broader the maintenance set, the more able the dog will be to cope with disruption of emotional homeostasis.

Disruption of Homeostasis

Disruption of emotional homeostasis can be caused losing one or more stimuli from the maintenance set, and leads to a feeling of reduced control. The extent to which disruption occurs increases with the importance of the lost stimulus or stimuli. For example, a puppy may whine all night after it is homed because the maintenance set it had built up at the breeders is totally disrupted when it is removed from its dam, littermates and the breeder's house or kennel. Another example is provided by dogs which become so dependent on their owner's presence that they lose behavioural organisation, manifested by an inability to stay in a positive emotional state, when left alone.

Things the dog finds frightening can also cause disruption of emotional homeostasis. These can be learned if a stimulus becomes associated with a threatening event, such as darkened skies becoming a predictor of fireworks. Dogs may also react fearfully towards a stimulus because of its physical characteristics without having to learn that it is threatening. Examples include intense and/or sudden stimuli, like gunshots, fireworks or thunderstorms. As behavioural organisation increases, the presence of novel stimuli and stimuli that do not perform to expectation may lead to a negative emotional state and loss of emotional homeostasis.

The Sensitive Period of Behavioural Organisation

The sensitive period of behavioural organisation can be divided into three stages.

INITIAL STAGE

A puppy's initial maintenance set normally consists of its mother, littermates and nest-site. This is unavoidable because of the availability and salience of the stimuli and absence of opportunity to associate other stimuli with a relaxed emotional state because puppies are born deaf and blind and can only crawl. Most of their time is spent sleeping or feeding. Although several primitive sensory and behavioural systems are present at birth, the puppy's behaviour is largely reflexive, and organised to guarantee what it needs for survival at this stage of development, such as food and adequate warmth. Neonatal puppies are incapable of controlling their body temperature internally but keep it constant by staying in close contact with their mother and littermates and they will display intense distress vocalisation when separated from them, but, provided they are warm, will be content to go to sleep without the comfort of contact with them.

Although complex learning cannot occur during the initial stage, simple conditioning of touch, taste and scent stimulus-response associations are possible from the first day of life. For example, Fox (1971) placed aniseed oil on the mammary glands of a nursing bitch. Puppies exposed to the smell whilst suckling were more attracted to it away from the suckling context than puppies which had not had the opportunity to develop the association.

Neonatal experiences seem to have considerable effects on the development of behaviour. This appears to have been illustrated in unpublished work by De Witt, which involved half litters of newborn puppies having no direct exposure to humans, whilst the remainder received a high level of human scent for half a minute immediately after birth. Subsequently both groups were isolated from human contact until tested, some at five weeks and some at seven weeks. Puppies which had been exposed to human scent had a preference for investigating people over other environmental stimuli. Unfortunately puppies which had no direct human contact showed the same preference, attributable to the fact that they were mixed with littermates from the exposed group, which would have been carrying human scent on their coats. However these combined groups showed significantly greater interest in investigating people when compared to a new control group which was kept in total isolation from human scent until three weeks of age.

A synthetic version of a maintenance stimulus called 'Dog Appeasing Pheromone' (DAP) is available as a plug-in diffuser. The appeasing pheromone is produced by the sebaceous glands in an area between the bitch's mammary glands three to five days after puppies are born. It helps the puppies to relax and will create a dependency upon her. So the resultant positive emotional state enables the puppy to remain relaxed as it explores its environment and encounters new stimuli in the next stage, and this reassuring property persists into adulthood. The use of a diffuser can provide a puppy with a salient maintenance stimulus when it is moved to its new home.

SECOND STAGE (THREE TO FIVE WEEKS OF AGE)

After the eyes open at around thirteen (+ three) days and ear canal opens at approximately eighteen to twenty days, a puppy becomes aware of visual and auditory stimuli. The puppy's movements also become increasingly coordinated and it develops the ability to run, but without the grace of an adult. The ongoing development of cognition, perception and motor abilities means that puppies experience a rapid increase in stimulation and, because complex learning is possible, most reflexive behaviour that was displayed automatically without conscious thought is replaced by learned responses related to environmental stimuli.

During the three-to-five-week period puppies are at an optimum stage of development for establishing a broad maintenance set. This occurs for several reasons: first they are less inclined to stay close to the maintenance stimuli they became dependent on in the initial stage (the dam, littermates etc.). In addition to their new mobility, their 'seeking system' within the brain stimulates them to search and explore their environment actively. Secondly, the brain overproduces neurons during early development that will go to waste if they don't get used. Although new ones are formed during learning in later life, the capacity to store knowledge of the world and how to cope with it is more efficient during these important weeks. Through exposure to their environment and interaction with it, puppies store representations of new stimuli and variations in familiar stimuli. Thirdly, during these weeks puppies are in a biologically pre-programmed period of parasympathetic dominance, characterised by factors associated with relaxation, such as low heart rate, predisposing them to have a positive emotional state. While they are in this relaxed frame of mind, puppies easily associate new stimuli with that emotional state of relaxation, and they become part of a broader maintenance set.

At the start of the three-to-five-week period, approach and investigative behaviour is directed equally to novel and familiar objects but most attention is paid to stimuli that change rapidly, such as movement and sounds. As recognition and recall memory develops and maintenance sets are refined attention to familiar things declines. Greater attention is paid to those things that are less familiar. These trigger investigative behaviour, whereas things that are extremely unfamiliar begin to cause some minor hesitation in approach. This can be explained by the fact that when something unfamiliar occurs the puppy's brain compares it to the representations it has stored. A successful match will result in a positive association and the further development of behavioural organisation. Failure to match during this phase of development will normally result in efforts to investigate the stimulus or event.

Preparedness to respond positively to novel stimuli during this stage was shown in a response test conducted as part of an experiment by Freedman *et al.* (1961). In this experiment litters of Cocker Spaniel and Beagle puppies were kept with their littermates and dam, but isolated from human company except for specific periods. Pups from each litter were taken from the field for a week of socialisation at two, three, seven and nine weeks of age and then returned to the field. During this week indoors pups were played with, tested, and cared for daily in three half-hour periods. How avoidance behaviour develops was evident in daily ten-minute tests of the amount of time a puppy spent in physical contact with a passive, reclining experimenter. The two-week-olds were too immature to do much but sleep, eat or crawl about randomly. However, the three-weeks-olds were immediately attracted and spent most of the ten-minute period pawing, mouthing, and biting the researcher and his garments; five-week-olds exhibited wariness at first, but they became comparable to three-week-olds before the end of the first play period. In contrast seven-week-old pups were frightened and wary of contacting the experimenter over the first two days, while nine-week-old pups exhibited these reactions over the first three days. Puppies which had not received any exposure to people were tested at fourteen weeks and were fearful, and remained so even after several weeks of petting.

Experiments have also been designed to reveal a puppy's response to inanimate objects, for example, puppies housed in conditions devoid of objects were placed in a test area with various articles for just half an hour at five, eight, twelve and sixteen weeks. These puppies were found to increase their exploration as they grew older and developed a preference for more complex things. However, puppies first tested when they were over eight weeks tended to withdraw from the items and those not tested until twelve or sixteen weeks old frequently became catatonic with fear (Fox, 1978).

During the three-to-five-week period dependency on maintenance stimuli for emotional homeostasis develops very rapidly and becomes progressively important for behavioural organisation. Search behaviour for missing social or non-social maintenance stimuli starts to occur, along with a gradual evaluation of contexts. In practice this means that to increase the chance of a puppy accepting unfamiliar stimuli without causing loss of emotional homeostasis, novel stimuli should be introduced in a familiar environment, so that the puppy can rely on its existing maintenance set. It also means that a puppy has to be exposed to a great amount and variety of animate and inanimate stimuli, and will benefit from being regularly moved to locations where unfamiliar types of stimulation and variations of familiar stimuli exist. This is at a time when the puppy is in the care of the breeder and therefore this responsibility falls to them.

THIRD STAGE (FIVE TO SEVEN WEEKS OF AGE)

After the rapid developmental changes in the brain, nervous system and behaviour, from five to six weeks development continues at a more gradual rate. The 'natural relaxed frame of mind' is lost and replaced by a period of high heart rate and sympathetic dominance from week five to a peak at seven/eight weeks followed by gradual decline.

The increasing maturation and integration of various brain structures during the five-to-seven-week period increases the potential for a reaction to novelty. Things the puppy has not had a chance to develop a positive emotional response to now have an increased likelihood of triggering fear, which may be displayed as either avoidance behaviour, fear or fear aggression.

Although the puppy has established a broader maintenance set and behavioural organisation in relation to the known environment, the composition of stimuli in it can change through the availability and importance of other things it encounters. In general, things that are moderately different or unpredictable will still cause exploration and a preference for investigating novel stimuli and locations over those that are familiar continues, so long as contextual factors are reassuring. However, the introduction of novelty becomes more likely to upset than to benefit the system. The amount and nature of previous experience increasingly determines whether the puppy shows fearful responses and an ability to control its environment, and thus maintain behavioural organisation. If a dog does not have a sense of control it may easily learn fears or develop what, from our perspective, may seem to be inappropriate and unpredictable behaviour in response to unfamiliar or aversive stimuli. Once developed these responses may generalise to other things and situations.

During this stage puppies should be given plenty of opportunity and time to become familiar with new things and build up their confidence to approach and investigate them. Giving them the chance to do this at their own pace allows them to control their emotional state, which they can exercise by governing their approach, withdrawal and wait and see responses. If they stay relaxed the stimulus will eventually be associated with a positive experience. Conversely, forcing a puppy to approach will have the opposite effect because a sense of loss of control will cause anxiety or fear, resulting in a negative association.

To maximise the chance of the puppy forming a suitable maintenance set and coping with the world beyond the breeder's environment, puppies should be homed as soon as it is allowed, which in the UK means from six weeks. There is some evidence to suggest that puppies may be particularly affected by unpleasant events at eight weeks, which is of special concern because this is when the majority of puppies are homed (Fox and Stelzner, 1966). The younger the puppy the easier it will be for it to add new stimuli it encounters to its maintenance set. After the age of seven to eight weeks, when the puppy's representation of the world is, for the most part, formed this becomes increasingly difficult, but definitely not impossible. Obviously, the opposite is true and the later the broadening of the maintenance set starts the less likelihood of developing sound temperament there is and starting from scratch at twelve to fourteen weeks is likely to present new owners with extreme difficulties. Of course, change is possible but prevention is better than cure. After they are obtained and have received their initial vaccinations, the process of exposing puppies to as many different stimuli and environments as possible to build up their confidence should continue. This process has to be repeated until at least six months and probably a year to prevent behavioural organisation regressing.

From 1959 practical application of the research into the effect of early experience was conducted with over 24,000 puppies by Guide Dogs for the Blind Association in the UK. This work is ongoing but was reported on in 1991. The qualities required of a potential guide dog, the training of which starts at around twelve months, include confidence and low reactivity to normal stimuli. Until 1956 the GDBA relied on the donation of adult dogs to maintain their stock and the success rate of those accepted into training and becoming guide dogs fluctuated between 9 and 11 per cent. In an attempt to improve the success rate puppies were purchased and placed in private homes between ten and twelve weeks of age or even later but the success rate did not significantly improve.

In 1959 the GDBA started managing the gene pool with its own breeding scheme. The puppies were raised in kennels and placed in the urban homes of puppy walkers after first vaccination at six weeks. Systematic exposure to the environment in the area around the puppy walker's home began immediately and continued until the puppies were returned for training. Annual success rates at between 75 and 80 per cent became common, with some variation between breeds. Since then the GDBA has also tried to ensure that their pupies are born and raised in a domestic environment rather than kennels.

Dogs which exhibit avoidance or aggressive behaviour related to fear should be more likely to have a history of limited early experience than dogs displaying other types of behaviour problem. This theory was tested by Appleby, Bradshaw & Casey in 2002 when the early experience of 820 dogs referred by veterinary surgeons to one pet behaviour counsellor (D. Appleby) were compared. The results suggest that there is a relationship between a dog's early environment and the development of fear-related behaviour problems. Living in a domestic environment at the breeders and experience of urban life after vaccination appears to reduce the probability that the dog will develop avoidance behaviour or aggression in response to unfamiliar people. This research also suggests that homing puppies raised in non-domestic environments before eight weeks also reduced the risk of these problems. Interestingly, in later life, there was a significant relationship between the dogs which had been reared in a non-domestic maternal environment, such as a barn shed or kennel and aggressive behaviour when examined by vets. A possible explanation is that for puppies which have not had the diversity of experience available in a domestic environment the first experience of visiting a vet, which normally occurs within two days of being obtained, is likely to be traumatic, and will result in an association between the vet, fear and defensive behaviour during later consultations.

HOME ALONE

The importance of exposing puppies to unfamiliar stimuli gradually also applies to teaching them to adapt to changes in their maintenance set when left alone. Elliot and Scott (1961) found that puppies which are not exposed to separation experiences early in life have an extreme reaction to them later on. They divided some puppies into four groups and exposed them on a weekly basis to separation in a strange pen. They repeated this until the age of twelve weeks. Puppies of group four were not exposed to separation experiences until they were tested for the first time at twelve weeks. These puppies panicked and were unable to cope whereas the other groups, especially the group that started having this experience at three weeks, had learned to adjust more effectively. Care should be taken to develop a puppy's capacity to cope with segregation gradually. Young puppies have a strong need for close social contact. This need is first satisfied by its mother and littermates and later by its owners. Locking a puppy in a room on its first nights in its new home, after its previously established maintenance set has been removed, and letting it cry until it is worn out is not good for its welfare. It could sensitise it so that it reacts negatively when left alone on future occasions, increasing the potential for separation problems such as destructiveness, vocalisations and elimination. New owners can gradually accustom their puppy to being on its own by getting it used to an indoor kennel and placing this next to their bed, and gradually moving the kennel away from the bedroom once the puppy has learnt to settle at night, or they can sleep near the puppy downstairs for the first few nights, gradually withdrawing their presence once the puppy has adapted to its new environment.

Summary

Maturation and integration of structures in the brain make more detailed information-processing possible, and this results in changes in behaviour and an increase in emotionality. Once the naturally predisposed period of parasympathetic dominance and positive emotional state has declined and the maintenance set has been formed, unfamiliar stimuli encountered may cause a negative emotional state. The characteristics of these stimuli and the characteristics of the maintenance set will determine the extent to which arousal and disruption of behaviour occurs. The more effective the maintenance set is the better able to cope with disruption of emotional homeostasis the dog will be. The presence of an effective maintenance set also increases the individual's confidence to explore and broaden it. It follows that a failure to develop an adequate maintenance set during the period of parasympathetic dominance, between three to five weeks and beyond, should have a detrimental effect on the development of subsequent behaviour, and there is an increase in the potential for behavioural disorders to develop.

3
Puppy Classes – A league of their own
Erica Peachey

Case Studies

JINGO

Jingo, a medium-sized crossbreed was referred to me for aggression towards other dogs. If on his lead, when he caught the sight of any dog, big or small, male or female, he would lunge forwards, barking and snarling. Although initially better off his lead, Jingo had now reached the stage where he would race up to another dog if he had the opportunity and immediately grab it by its neck and pin it down. If the other dog fought back, a fight would ensue.

'I don't understand it,' Mrs Smith told me. 'He used to love other dogs. Everyone always laughed at the puppy classes because he never listened to me, he just wanted to play with the other dogs.'

Jingo had been acquired as a young puppy from a local rescue kennels. Keen to do the right thing, Mrs Smith and her family had enrolled him for puppy classes as soon as he was clear from his vaccinations. Jingo only had eyes for the other dogs: the highlight of his lesson was when all the puppies were let off the lead to 'socialise' with each other. He became the joker of the class, leaping over other puppies, bouncing on them, and generally encouraging them to play. How could a puppy who loved other dogs develop such an aggressive reaction, or was it this same rough play that had caused the aggression?

MILLY

Milly loves people. She has a lovely nature – when she sees people, all she wants to do is be close to them. Unfortunately, this means that her owner can no longer let her off the lead on a walk, because Milly will run over to anyone and jump up repeatedly. A powerful, young Golden Retriever, she has knocked young children over and has even knocked dog-loving adults off balance.

In the home, Milly's passion for people means that visitors are rare and the children are not able to have any friends to play unless she is shut away.

'The instructor at the puppy classes said that we should socialise her, so we took her everywhere and she met hundreds of people. She was so cute and small then, that no one minded when she jumped up. Some people even encouraged it.'

Was Milly's owner given wrong, or incomplete advice? Would Milly have been a better behaved dog if she had not attended classes?

SHEEBA

Miss Brown told me that Sheeba had always been the perfect dog. Sheeba had always been shy of other people and dogs and on walks she did not cause any problems, keeping her distance from everyone else and never moving far away from her owner. Then one day, when Sheeba was five years old, she was on her lead on the way to the park, when a small boy jumped out from his drive and threw his arms around her. Sheeba struggled to get away, with a yelp or a snap, but unfortunately a tooth grazed the boy's cheek. It was not a serious injury, and fortunately the boy's parents were understanding, but Sheeba's owner was devastated. Sheeba had not attended any puppy or dog classes. 'I never saw the need,' Miss Brown told me. 'She always did everything I asked of her.' Would Sheeba's attendance at puppy classes have meant that her negative response to this incident would have been avoided?

There is no easy answer to these questions. We cannot know because we are dealing with pet dogs in families, and therefore cannot run controlled studies to find out what would have happened, had things been different. However, certain patterns are identifiable and this allows us to understand the causes and what programmes of behaviour modification are likely to help.

First Experiences

We all remember our first experiences of something important. Music lovers will know the first record they ever bought, but probably do not know the second. I remember how much a bag of chips used to cost when I first bought one, but have no idea when they went up. How many people remember their first kiss? But what about the second? First experiences are very powerful in our memory and there is no reason to expect this to be any different in the dog. So it is essential to ensure that the first time creates the right habit and the correct learning. In times of stress many of us revert to the first thing we were taught, often without realising it.

Therefore, if a puppy's first experiences were unpleasant, resolving what it learned can be a long and difficult process. However, good initial experiences will greatly assist a dog throughout its life.

Socialisation and Socialising?

There are many and varied definitions of socialisation and these change according to time and context. We talk about adults 'socialising' during a night out, children 'socialise' when they play together. According to different dictionaries, 'socialise' can mean: 'to behave or interact in a sociable manner', 'to make somebody fit or prepared for life in society' or 'to make social', 'to act in a sociable manner'. So, in general terms, socialising refers to behaving in a sociable way, or the process of learning to do this.

However, 'socialisation' has a stricter ethological meaning. It can be defined as 'the process during which animals are particularly receptive to stimuli and when various social preferences are established.'

It is commonly thought of as the process of learning how to recognise and react correctly with the species with which it is familiar. For wild animals, such as a wild dog or wolf, this is relatively easy as this involves one species. However, for our pet dogs, at least two species and possibly more are familiar – humans and dogs – and possibly cats or others. Some regard this as socialisation, others do not. We can immediately see that we are expecting more of our pets and that selective breeding has allowed them to cope with this 'extended family' as their wild counterparts never could.

Another definition of socialisation is that 'an animal develops appropriate social behaviour'. So a dog is said to be socialised to something, when it displays a recognition and acceptance of it and behaves accordingly. It therefore follows that two puppies may experience the same circumstances but, because of differences within them, one may be said to be 'well socialised' and one may be 'poorly socialised'.

Because of the popularity of pet dogs, the definition of socialisation has broadened in meaning. Rather than simply referring to the development of an understanding of one or two species, it is commonly used to mean the entire process of reacting to and fitting in with life as a pet dog. In the past, this has been termed "social referencing" but as discussed in the second chapter – the development of maintenance stimuli to enable behavioural organisation and emotional homeostasis is a more accurate concept. This model is able to cover social and non-social stimuli, i.e. learning to adjust with people, traffic, vacuum-cleaners, etc. The aim is to equip the puppy or dog with the necessary knowledge to allow it to adjust and behave in an acceptable way in our society. It results in a dog which is better able to cope with novel experiences, unpredictable situations and fearful circumstances. Breed differences play a part in this. Some breeds and types have been selectively bred to react in a certain way. For example, gundogs are normally required to work closely with people and react well to other people and dogs. Livestock-guarding dogs are bred to bond closely with their flock and keep intruders away. Terriers have been required to work independently of their owners, pursuing prey and not back down from threats.

So What Does This Mean in Practice?

We want to raise puppies which can cope well in our society. What we are aiming to achieve can be described as: '*an active process of teaching a dog how to behave in all situations, through pleasant experiences.*' This definition encompasses several important points.

- '*Active process*'. This means actively searching for different situations and controlling the circumstances to ensure that the right outcome is achieved rather than waiting for things to happen.
- '*Teaching*'. We expect a great deal from our dogs and want them to behave in ways that they would not normally do. So it is unrealistic to expect them to learn it all by themselves. As the humans in the relationship, we must do the teaching.
- '*A dog*'. This applies to a dog throughout his life. Although puppies learn more quickly and have a greater need to learn, dogs are extremely adaptable; dogs are good at learning, and this enables them to be kept as pets. Dogs of any age can learn, but prior learning must be taken into account.
- '*All situations*'. All the circumstances in which a pet dog may find itself. Thought and consideration must be given by owners and those involved in puppy education about what situations can and should be included.
- '*Pleasant experiences*'. This term reflects the fact that dog training has moved on considerably in the last twenty years. The old style of force and punishment is no longer acceptable and has proved to be unnecessary. Deeper understanding of dogs, relationships and learning have meant that reward-based methods that are kind, fair and effective are more commonly used.

There is no set time scale for this process because learning is ongoing. Things that can be learned can usually be unlearned and a dog's behaviour will usually change over time. Consequently, active social experiences must form a regular part of life, especially if social behaviour was not well developed in early life.

A litter of wolf puppies which were hand raised, but still had access to other wolves, gradually reverted to their natural suspicious behaviour and were wary of humans even though exposure to humans was constant. As they matured, they chose to keep away from the humans, even those who had hand-fed them. A litter that was hand raised away from the company of other wolves, learned to enjoy people being around and were able to be reintroduced to the pack and still retain their enjoyment of being with people.

Where Do Puppy Classes Fit into This?

Puppy classes have developed from a tried and tested idea brought to the UK by Ian Dunbar in the 1980s. Although puppy classes had existed previously, the idea of a class full of young puppies, just after their vaccinations, was a new one. However, like all great ideas, the general theory was sometimes misunderstood or misapplied. Puppy classes – i.e., a class for young puppies to mix and learn through reward-oriented methods – gained much popularity and interest, and others tried to emulate it. A puppy class can simply mean a beginners' class, where young puppies are alongside bouncy adolescents and maybe older dogs, perhaps showing anti-social behaviour. Exposing a puppy to these conditions can cause problems.

The term 'puppy class' does not specify an approach to training. 'Old-style dog training', of yank and pull, was originally only for dogs of six months or older. This was because it was thought that puppies under six months were not able to cope with the stresses of this type of training. However, using motivation, lures and fun meant that puppies could begin training from the moment they came into their new homes, thus removing the need for this restriction. Unfortunately, when the 'old-style training clubs' sensed competition from forward-thinking trainers, they too reduced their age limit, thus exposing young puppies to choke-chain and force.

Everybody has a slightly different concept of a puppy class. It is a good idea for an owner to consider what they want, using this chapter as a guide, and invest time in finding the right class for them. Each puppy will also have different needs. A good instructor should recognise what they are and cater for them, by providing controlled situations, advice and support for the owner. The nearest puppy class is not necessarily the best, neither is the cheapest. Find out what is available, speak to your veterinary surgeon and other dog owners, talk to the instructor on the telephone and perhaps visit the class to watch, without taking your puppy, preferably before you obtain it.

A puppy class should differ from many conventional training classes in several ways:

- It should be less structured or regimented. Puppy groups should be small enough to allow the instructor to be familiar, with each puppy and family and adapt the course according to their needs.
- It should give more freedom and play which can then be developed into owner control. Teaching about games and the importance of playing the right ones should form a part of classes for any age, but practical exploration of games is possible with puppies.
- It should include some off-lead sessions (see later). This does not mean off-lead 'free-for-alls', rather structured sessions which involve off-lead work, such

as coming when called, calling away from other puppies, play etc.

- It should provide exercises that are varied and imaginative, teaching what is relevant to a pet dog rather than obedience-style routines. For example, teaching a puppy to learn to come when his owner calls, and leave another puppy it was sniffing, rather than simply return to his owner after being positioned sitting and facing the owner, anticipating the next part of the exercise.

There is a difference in the attitude of owners attending puppy classes. Rather than attending class out of desperation with an adolescent that is causing increasing chaos, owners bring their puppy, who is special to them. Although he may not be perfect, owners are normally far more positive than they may be if they attend classes for the first time when the pet is older. Owners are also at their most motivated and receptive and, because results tend to come quickly, they usually try harder.

Puppy classes can only be as good as the instructor who takes them. The term itself does not guarantee a class which will be an ideal learning environment for any puppy. When I use the term for the rest of this chapter, I will be talking about properly run classes where numbers and ages are limited and the instructor has a good working knowledge of owners' and puppies' learning and motivation.

Some classes are termed 'Puppy Socialisation Classes'. Whatever definition is used, this is somewhat restricting the scope of a class. I feel that puppy classes should actively teach:

- good manners
- how to respond to their owner's wishes

- responsible dog ownership
- how to avoid problems
- how to overcome typical puppy difficulties, such as mouthing, and any house-training problems,
- how to increase the owner's understanding of their puppy
- how to shape their puppy's confidence and behaviour in all situations.

The title 'Puppy Classes' more aptly covers this.

Whilst on the subject of terminology, many owners tell me that they and their dog or puppy have taken part in 'The Socialisation Exercise' at local dog clubs. For those not familiar with it, the exercise usually consists of young puppies or beginner dogs or, more worryingly, a mixture of both positioned in a circle or line. One at a time, one owner and puppy weaves in and out of the other dogs, usually accompanied by much shouting of 'Leave it!' and pulling from all the owners. How can this be teaching dogs to be sociable towards each other? It seems far more of an 'anti-socialisation' exercise. From the dog's point of view, stress is inevitable, either from the other dogs being close, possible aggression from other dogs, from not being able to behave in a natural way, or from the noise and action of the owners. If we believe that the aim of puppy classes is to increase a puppy's ability to cope, and to teach it how to behave in normal settings, it is hard to see how this exercise is anything other than destructive.

What Should Be Included in a Puppy Class?

A good puppy class should be flexible and therefore the content of the course should vary to some extent. Some areas should always be covered, such as puppies learning to respond to their owners, owner understanding, social manners and avoiding problems. Within this, there should be room to cater for the individual needs of the puppies and owners.

Some owners imagine that all the social experiences their puppy needs should be provided by the puppy class. Although a puppy should gain valuable experience, it is unrealistic to think that this is sufficient. Dogs learn according to the situation, so even if a puppy does experience a wide range of people – men with beards, women with hats, children with sunglasses, disabled adults, people from different ethnic groups, etc. – there is evidence to suggest that this may only enable the puppy to accept all of them within the confines of the classes. In order for the growing puppy to behave appropriately in all circumstances their education must be continued at home and on walks. Therefore, the role of the classes is to:

- provide basic social experience, i.e., good experiences with different men, women and hopefully children of all ages as well as other puppies
- explain to the owners the importance of teaching acceptable behaviour, and encourage them to continue and broaden the education between classes
- note the reactions of the individual puppies and give each owner specific advice, including areas to concentrate on, times to take extra care and how to avoid problems.

What Else Should Be Included?

FREE PLAY

This is an area of some controversy. The original principles adopted from the US involved large amounts of free play for the puppies. However, in practice, in this country, it was found that many puppies were actually learning to fight during these 'play' sessions. Some puppies played so roughly that aggression became part of their game. This was especially true of some breeds such as terriers. Others became frustrated when their expectations of rough play were not met, which, in later life, can result in aggression. Some puppies were overwhelmed by their boisterous playmates and learned that aggression was the best method of defence. Owners likely to attend a puppy class are also the owners who take their puppies out and about on walks, on which most will meet and interact with other dogs. As a consequence, they are already learning fluency in 'dog language' and this, combined with the artificial, relatively sterile environment of a puppy class, i.e. where there are fewer distractions and the owners are unlikely to disappear by walking away), means that games become the only thing to concentrate on, and therefore, frequently get out of hand.

In a good puppy class, each puppy will be assessed and the course tailored according to its needs. A shy puppy may be off the lead and encouraged to interact and have good experiences with others, whereas the owner of a boisterous puppy may be shown how to be more interesting than everything else, by using rewards and motivation. The puppies are individuals and different exercises are needed to bring out the best in each of them. This approach must be explained to the owners, to develop their understanding and to ensure that no one feels 'short changed', or that favouritism is leading to different treatment.

Free play sessions, although a superficially easy option for the instructor, are not recommended other than in exceptional circumstances. However, off-lead sessions, when one, two or more puppies are off their leads whilst working with their owners, are achievable in most puppy classes and can be extremely beneficial.

RESTRICTION AND RESTRAINT

Learning how to behave when restricted is very important for puppies and something which is often overlooked by owners and instructors. If puppies are used to freedom and the ability to interact as they choose, being restricted on a lead is contrary to their expectations and they may learn to object. A gradual introduction to being restricted on a lead is an important step.

APPROPRIATE BEHAVIOUR WITH ADULTS

Teaching social behaviour involves teaching appropriate behaviour. Teaching acceptable behaviour varies between puppies and age groups. A puppy that is nervous of people needs to learn that people are linked with good experiences and that being around them is enjoyable. A boisterous, extrovert puppy which adores everyone does not need to learn to enjoy being with people. It already knows this, but it needs to learn to greet them in an acceptable way. It needs to learn that jumping up results in no attention and keeping feet on the floor is the best way of seeking attention, although this does not guarantee that attention will follow.

MANNERS WITH OTHER DOGS

This must not be overlooked. Puppies will develop their own rules for games and interaction with older dogs which is essential for well-balanced behaviour. A puppy should meet playful dogs as well as dogs that are less inclined to play. They should meet dogs which want to be with them and dogs which do not. Obviously, a puppy should not be around a dog which is aggressive towards puppies, but understanding that not all dogs want to play is an important skill for a puppy to learn. If a puppy does not learn to read dog language it is likely to encounter problems as it matures. This should not necessarily be provided during the puppy classes, as finding suitable adult dogs and ensuring that they are not stressed is very difficult, but the process must be explained to owners, along with help and encouragement given at each session.

UNDERSTANDING CHILDREN

Manners with, and a basic understanding of, children is something that all puppies should learn. Learning may be limited in a class environment, but it is important to make the effort, perhaps by having one week as 'Children Week', where everyone is asked to bring children if possible, and asking children who attend a different training class to come and meet the puppies. It must be explained to owners that this is not the end of the process and that it is one that must be followed during the weeks ahead. If a puppy lives in a household with no children, every effort should be made to find as wide a variety as possible. If there are children in the family the puppy usually understands them, but children of different ages should be introduced.

HANDS ON

In the old days of forceful training, most dogs became used to rough handling. Although the means did not justify the end, this tolerance was a useful by-product of the harsh methods. Modern methods, although effective, enjoyable and largely stress-free for puppy and owner, do mean that a puppy receives less hands-on contact. This can cause problems with grooming, administering tablets, being handled by a veterinary surgeon and sometimes even being stroked by a stranger. Therefore, as well as the 'hands free' methods of training which involve luring rather than forcing, a good puppy class should teach the importance of lots of handling, throughout its life. During the class, this can take the form of physical praise, handling and grooming by the owners and others, with many people touching the puppy in a pleasant way. As with all other aspects of the classes suggestions for practising at home and recommendations for each puppy as an individual are what make the difference between an average class and an outstanding one.

What Does the Owner Need?

Having looked at what the puppies need from a class, we must also consider what the owner requires. This can largely be divided into five main areas:

- *Information and Education.* This is obvious, but the class should be small enough to allow for specific as well as general points to be included.
- *Acceptance* is important. No one likes to be told off and, as adults, very few us of are prepared to give up our money and time for this. Many people still believe in the outdated concept of 'No bad dogs' (remember first learning experiences and the impact they have). This means that the class becomes a daunting place if the owners feel they will be blamed for their puppy's less than perfect behaviour.
- *Support* is reassuring and essential. Some of this comes from the instructor, sometimes it can come from other owners. For example, when the instructor says that

mouthing is natural for puppies of this age, it may help, but when another owner confides that their puppy does the behaviour it can put any worries into perspective.

- *Motivation* of adults can be difficult. However much they love their puppies and have chosen to come to classes, we all lead busy lives which sometimes verge on hectic. Motivating adults to change their behaviour and to spend time on walking, socialising, playing and teaching can be difficult. Consequently the instructor needs to understand human, as well as canine psychology.
- *Enabling* is vital. 'Give a man a fish and you feed him for a day, teach him to fish and you feed him for life.' Enabling an owner to lure his puppy into a 'sit' position is pleasing for the owner, but teach him how to teach his puppy how to respond to him, and he can teach that puppy anything he chooses. A puppy class only runs for a few weeks, but the skills an owner learns should stand him in good stead for the entire life of his dog.

There are many other aspects of puppy ownership which can, and perhaps should, be included, such as responsible dog ownership. This is not a comprehensive list of what a puppy course should include. For many years there has been an emphasis on teaching response to commands, i.e., a puppy must be able to sit, stand, lie down and stay. This is not included in most people's requirements. Practical exercises such as coming when called in a natural setting, walking on a loose lead and greeting people without jumping up are much more likely to be listed.

Puppy Parties

Puppy parties can greatly benefit a puppy, owner, and their veterinary practice. The definition can vary, but a 'puppy party' should mean a group of very young puppies and their owners being invited to a veterinary surgery for a one-off meeting or a short series of meetings. The puppies are normally very young, usually between first and second vaccination. The group should be small, up to about eight puppies, of different breeds and sizes but of similar age. Whole families should attend, providing a variety of learning experiences for the puppies. During a puppy party, the puppies and owners receive an introduction to the veterinary practice and form pleasant associations with it. They are also introduced to the concept of socialising and other behavioural organisation and to training and understanding their dogs. Avoiding problems and resolving common difficulties such as mouthing, lack of house training etc should also be included. As the puppies are so young, more 'off-lead' time is usually appropriate. This is an age when puppies are quite often isolated from many of the situations, and experiences which they so desperately need because of the vaccination process. A puppy party goes some way towards compensating for this confinement. Owners are usually extremely receptive at this stage and therefore can enjoy the puppy party and be motivated to continue to provide social experiences and learning. There are risks to puppies mixing before their vaccination programme is completed, but at puppy parties these are minimised and therefore, it is as safe as possible. Although it is hard work for the veterinary practice to organise and run, the puppy party yields huge benefits for the practice and its clients.

Conclusion

So where do we go from here? It depends on what we are looking for. For those involved in research, there is a wealth of opportunity to study puppies in pet homes and to monitor the depth of quality experience and its effects. For those involved in puppy education, we constantly strive for improvement. With the advances that have occurred in training methods, instructor courses and support, knowledge is now easier to find but may be harder to keep up to date with and to put into perspective, so that it can be made a useful tool. For those with puppies, decide what you want and find ways to achieve it, by 'wearing your puppy' and taking him out and about with you. Hopefully, a good puppy class will help you achieve your aims, understand more about your puppy and have fun in the process.

Think back to the three dogs we met at the beginning of the chapter. Have your views on them altered in any way?

4
Pets and Children

Erica Peachey, Donna Brander and Emma Magnus

Children and Dogs

Erica Peachey

Introduction

Pets and children together is a happy ideal that we have in our minds. Whether it is on old-fashioned chocolate-boxes, in Lassie films or advertisements for toilet rolls, the idea of adorable children and cute animals can be a far cry from reality.

For both pets and children, being involved with each other has tremendous advantages and is worth the effort, but, is undoubtedly hard work for the adults involved. Studies have shown that children raised in households with pets tend to relate more closely to others, have greater empathy, increased social skills and may have a better understanding of life's difficulties. For pets, especially dogs and cats, the rewards can include a richer, more varied lifestyle, increased companionship and time spent with them. Sometimes, I wonder why I have found no studies outlining the benefits to parents. On difficult days with my family, I can imagine why this is, but most of the time, the rewards are obvious.

Dogs pose the biggest potential for problems with children but also, perhaps, the greatest opportunity for good times, so the majority of this chapter will focus on them. It is divided into four sections. The first looks at dogs and children in the same household, growing up together with all the difficulties this entails. The second section looks at some of the special considerations which may be needed. Thirdly, we look at advice for dog owners, and, finally, advice for parents.

Know your own dog and/or child and follow, adapt or disregard the following advice accordingly. If you have any concerns regarding the behaviour of a pet, speak to your veterinary surgeon and ask for a referral to a reputable behaviour counsellor, such as a member of the APBC.

Getting the Best out of the Relationship

PROVIDING GOOD EXPERIENCES FOR BOTH CHILD AND PET

What constitutes a 'good experience' will vary between different children and different pets. For dogs, this will tend to include walks, games, food treats and attention. It is preferable that the dog has learned good manners and the boundaries of acceptable behaviour before being with children. For example, if a dog snatches treats, barks constantly for toys, jumps up when going for a walk, etc, pet ownership is unlikely to be a good experience for the child involved.

REDUCING OR STOPPING BAD EXPERIENCES

This may involve:

- Limiting the time the animal is with your child. This is especially important if your child is active or excited.

- Not allowing your child to chase a dog which does not want to play. What is the child learning? What is the dog learning? How can they learn to like each other's company from that experience?
- Not allowing children to tease/hurt/pester a pet. This may sound obvious, but it is amazing how often it is overlooked.
- Not letting a pet frighten your child. Barking, jumping up and licking may all be perceived as friendly, but can be very intimidating for a young child. Therefore, take special care with regard to greeting and games, and especially whom puppies which have not learned to stop mouthing are involved.

OTHER POINTS

A child's presence should signal good things. Find ways of including the child in the pet's favourite activities. For example:

- walks with the baby and Rover,
- games with the child involved or sitting on a knee,
- cuddles with everyone involved etc.

It is tempting to wait until the child has gone to bed and then have quality time with Rover, but what does the pet learn? That life is great as soon as the child is out of the way? Added to which, without interacting with the pet,the child cannot learn. While some things are better done when the children are not around, ensure that good things happen when they are together. When the child goes out or to bed, have a period of time of no interaction with the pet, so it does not learn that no child equals fun and attention.

Routine. Routine is often helpful for a pet, child and adult, but flexibility is also important.

THE THREE 'S's.
Safety. Safety must always come first. Every parent and every owner must be responsible for their own child/pet. If a situation leads anyone to feel uncomfortable, stop, separate and review it.
Supervision. Supervision varies. For example: the advice is that children should be supervised when in the bath. However, this level of supervision varies, from a new-born, which has to be held and constantly watched, to being in the same room, to being in vocal contact, until the child is old enough to be left alone when in the bath. Therefore, merely being in the same room may not be an adequate level of supervision, especially in the early stages of resolving a problem.
Common Sense. It has been said that the problem with common sense is that it is not all that common. Lots of information is available, especially as many people now have access to the internet. However, not all information is correct or applicable in each situation, so use your own best judgement at all times and seek qualified professional help where necessary.

Dogs and children living together

WHEN A DOG IS ALREADY IN THE FAMILY AND A BABY ARRIVES
In some ways, this is one of the easier situations, as there is normally a few months in which to make any changes. Thought and consideration are needed to ensure that the whole situation remains harmonious. A dog which walks nicely on the lead, does not jump up or demand attention and that comes when called is greatly appreciated when you have a young baby to consider. Before the arrival of the baby any behaviour problems and training

difficulties must be resolved. A behaviour modification course supervised by a behaviour consultant and/or a dog training course are better undertaken now, while there is more time and fewer complications. A thorough check-up and chat with your veterinary surgeon can put your mind at rest concerning any of your pet's health worries (see below).

BEFORE THE ARRIVAL OF A BABY

Decide what changes will need to be made, make them and stick to them. For example, if Rover will not be allowed on the furniture once the baby is born, make these changes as early as possible, so that the pet is used to them before the baby arrives. Not only does this mean that changes are made gradually, but the pet is less likely to view the baby as a limiting presence. It is also important to accustom the dog to the sights, smells and routines that will occur when the baby arrives. This can be achieved through making a tape of a baby crying, introducing scent impregnated cloths and practising the routines associated with baby care, perhaps with a large doll. Your midwife may be able to help you obtain tapes and cloths from other parents.

PROOF FOR ROUGHER HANDLING

Children can be heavy handed and should be taught to avoid hurting others. However gentle a child and however carefully you supervise, a child will behave inconsistently, touch in different ways and generally prove to be something of a challenge for your pet to cope with. Try to anticipate each stage of your child's development, and prepare your dog for it. Grabbing fur, hugging, tripping over, shouting, screaming, banging doors, pointing to ears, eyes, staring, jumping off furniture, etc., all form part of most children's repertoire at some stage, and so help your pet get used to these. For example, we want Rover to feel that it is perfectly normal for people to throw themselves on the floor, so that when the child does it, it is not an automatic signal to play. At a time when your child is not around (so he or she cannot copy), do a minor version of some of the things your child may do. Reward Rover for no reaction. Then do it slightly more, and reward Rover again and so on.

Children copy adults, so ensure that the way you behave with your pets is something you want your child to emulate. For example, if you frequently shout at Rover and tell him he is a 'bad dog', do not be surprised if your child reacts to him the same way. Even if it is a pleasant action such as patting, their version of it may not be enjoyed in quite the same way. Try to stay one step ahead by thinking of the things your dog will have to cope with and prepare it for them.

Consider any additional points that your pet may need to learn. Acceptance of being occasionally shut out of a room is essential, in my opinion. Teaching a dog not to touch toys left on the floor and to ignore the sound of squeaky and rattly toys by using positive reinforcement is also a good plan.

Develop a routine. It is tempting to spend all your time with the dog, especially during maternity leave, but this may not be the best for the dog because it may develop the wrong expectations that will be at odds with reality after the baby arrives. Exercise and games should continue, with as little disruption as possible. On a practical note it may be beneficial to speak to a professional groomer, even if your dog does not normally visit one. Having a dog bathed and excess hair trimmed can be a bonus. Poop-scooping can become more difficult for pregnant ladies, but is still an essential part of walking a dog.

Decide on what provision will be made for your dog when you go into hospital. Most new mothers and babies stay in hospital for a few days after the birth and it is a good idea for someone to bring some items of clothing home for the dog to smell before washing. The clothes just have to be left lying around for the scents to become familiar.

On returning home, it is usually a good idea for the mum to leave her baby outside (with someone to supervise) whilst she enters the home and says 'hello' to the dog. Then the baby can be carried in (car seats are wonderful for this) and the dog can say his 'hello' in whatever way is appropriate. If you are concerned about the reaction of a boisterous dog, attach a trailing lead to the dog's non-slip, e.g., buckled collar, and this way you can interrupt any over-friendly behaviour as well as rewarding calmness.

New babies are relatively easy for most dogs to accept. If there are going to be major difficulties, they are more likely to occur when the baby becomes mobile. Suddenly, this strange creature is moving and unpredictable. At this stage, protect the dog to whatever extent is necessary and begin to teach the child. Although they are extremely young and cannot retain all information, they are learning huge amounts and need to develop good habits. Good experiences of dog and baby together are essential, and this normally involves walks, games, titbits, etc., and may be cuddle times for parent, child and dog. Ensure that the dog has time away from the child and can always get to a safe area where it will not be followed by the child; baby gates, play pens and indoor kennels can all be useful here, if used correctly. Meal times can be difficult. I find that a good general rule is that dogs and children are fed separately and do not mix at these times.

WHEN A PUPPY IS ADDED TO A FAMILY

Any difficulties can be greatly reduced by careful thought and consideration regarding the choice of puppy including breed characteristics. Ensure that the puppy (or older dog) is acquired from a reputable breeder or home. I would recommend a puppy with an experience of home life, preferably with children in his early weeks, as a pet for a family.

If acquiring an older dog, ask why the dog needs a new home. Even if the situation is not due to behaviour problems, for example, a breeder selling on older puppies, I would hesitate to take on an older dog which has not been successfully integrated with children. This does not necessarily mean living with them, but the dog should have been with children frequently and enjoyed the experience.

All the points discussed in respect of introducing a baby apply, such as good experiences together, separating at times and teaching all members of the family, pet included, how to behave with each other. Good puppy or dog classes will be happy for the whole family to attend to develop this learning, depending on the ages, number and behaviour of children.

Special Considerations

- When a baby arrives, some older dogs cope with the family changes extremely well, enjoying the extra interest that life with a baby brings. Older dogs are generally well behaved and usually need less exercise, socialising and training than when younger. However, extra care may be needed as the older dog may be fixed in a routine and its expectations may take time to change. Also an older dog may find it difficult to get out of the way of boisterous toddlers and may find rougher handling painful and therefore need more protection from the child's attentions.

- It is likely that a younger dog is more demanding of your time, as it needs greater exercise and stimulation. The advantage is that it can adapt easily, but you must still give him the time he needs for stimulation, exercise, training and social learning.
- Children with special needs can benefit from being around pets but obviously especial care must be taken that the pet is suitable. There are no hard and fast rules as every situation varies, but always see the situation rationally and from everyone's viewpoint, including the dog's. Take advice where necessary and do not rush into making decisions.
- Where children regularly visit a dog, such as is frequently the case with grandparents or other relatives who own a dog, the situation must be approached with care. In some ways, this is easier as dog doesn't have to live with the child. However, it can be more difficult because the dog doesn't have the same opportunities to learn as a dog which a child lives with. The frequency and duration of visits are important factors. At one extreme is the close family, where the child and adults call in regularly, possibly unannounced, for varying lengths of time. In many ways, this is good as the dog is used to the child at every stage of development and the

child is familiar with it. However, if the dog is not happy with the situation, problems can arise. The same considerations apply as those for a baby, i.e. supervision, limiting the time together and ensuring both have good experiences.

At the other extreme is the family who visit and stay infrequently. They typically live further away and therefore, their visit usually involves one or more overnight stays and a disruption of routine. This can be more difficult, as the dog may only see the child every six or twelve months and so cannot be easily accustomed to the changes. The child too has not seen the dog and so, has less idea of how to behave with it during these visits. If dog and child really cannot mix, segregation or use of good boarding kennels is an option.

- The pet must not become a scapegoat for family tensions. Families are rarely harmonious all the time, but if there are unresolved issues, the pet should not be involved.

Meeting Unknown Children – Advice for Dog Owners

In these days when people readily turn to litigation, children can strike fear into the heart of a dog owner. What can be done to reduce the likelihood of any problems?

Socialising the dog is vital, and this should continue throughout a dog's life. Although patterns of social behaviour become established over the first few months of life, dogs are always learning and like a person driving a car, it is better to practise these skills on a regular basis. The concept of behavioural organisation (as discussed in Chapter 2) in part involves learning the right behaviour through pleasant experiences. Therefore, socialising to children should not be a matter of leaving a puppy with some children and hoping they all learn to get on or mixing with a couple of children and hoping this is sufficient. It is an active process that involves seeking out children of different ages and types and ensuring that your dog has a good time with them.

Socialising with children can be difficult, especially if the dog does not live in a household where there are children. Good puppy and dog classes provide the presence of some children but more will be needed. Walks in different places, mixing with friend's families and walking in busy areas in a controlled way can all help to increase the amount of contact with children. Be ready to make the most of any situation. For example, if a child or a parent shows any interest in your dog, ask if they would like to make friends.

Anticipate and avoid difficult situations. If you feel unsure of the child's or your dog's behaviour, walk away from situations you are in, removing your dog from potential trouble. This may be because a child is becoming too boisterous, a small number of children has become a large group or because you simply notice signals from your dog that it is no longer enjoying the contact with the child or children and is beginning to find it difficult.

Some children may be scared of dogs in general or of your dog in particular. Children who are scared can behave in a way which many dogs find unnerving, e.g., if the child tends to stare, squeal, move quickly, comes close and then moves back. Know your own dog and do not allow a situation to develop that it cannot cope with.

Other children can be too rough. These tend to be those who are unused to dogs and therefore have no idea of how to behave around them, or children who live with a very docile and tolerant dog and have learned that the way to interact with dogs is to pull their ears, hang from their necks, etc. Again, remain in control of the situation and be ready to interrupt them and give some guidance, or move your dog away if needed.

The attitude of parents is also worth mentioning. It came as a great shock to me when I realised how many adults are to a greater or lesser extent afraid of dogs: this can have a negative effect on their children. Make sure that you don't force your dog on anyone and just as you and your dog may need to 'escape' when you choose, the parent and child must have the same opportunity. A common misconception is that only aggressive dogs cause problems. This is not the case – many problems arise with friendly dogs, especially when they are off-lead. A dog which loves children can be an extremely frightening creature to a child who is not confident around strange dogs. Being knocked over by a dog is very frightening and many playful dogs become extremely excited if this happens. Take extra care if you feel this may apply to your dog. You may know that your dog is friendly, but behaviours that a dog owner may view as normal can be enough to arouse feelings of fear in parents and/or children. This will include barking, running to greet people, jumping up, being playful, running quickly, licking (especially faces), puppy mouthing etc. If your dog has a tendency towards any of these, you are likely to encounter hostility in public places but appropriate training and control can prevent these problems.

Always remember that your perception of your dog may not be the same as other people's. Many children are frightened of large dogs, but assume all small dogs are puppies and therefore want to be cuddled. Certain types arouse feelings in adults. As a generalisation, large dogs, black dogs, German Shepherds, Rottweilers and Staffordshire types appear to be viewed with greater suspicion and less tolerance than many other breeds. For example, owners of black Labradors tell me they meet more anti-dog feeling than is reported by owners of yellow Labradors. Due to reduced tolerance the presence of dogs around schools and children's play areas is increasingly frowned upon. This is such a shame because responsible dog owners and well-behaved, friendly dogs can be a source of great interest, education and fun for children, some of whom may not otherwise have much contact with them. When out and about with your dog in areas populated by children always take extra care and ensure that your dog, whilst learning his own social skills, helps others to view dogs in a positive way. One final word on being an ambassador for dogs: never forget your poop scoop. Carry it with you and use it when necessary.

In summary, be a responsible dog owner. Each time you are out you are contributing to people's view of dogs. Make sure that no one has cause to fear or worry about your dog or its actions and, if possible, let children meet him and learn a little more about the enjoyment of dogs. Socialise your dog with children and be one step ahead in each situation, anticipating any possibility of problems and ensuring it does not happen.

Meeting Unknown Pets – Advice for Parents

No one wants their child to be hurt, and being bitten by a dog is a source of fear for many parents. Usually this fear is based on lack of knowledge, or perhaps some experience in their childhood.

Teaching children to be safe around unfamiliar dogs is as important as teaching them how to behave around cars. What we teach and what we expect depends on the age of the child and therefore learning must be developed as the child grows. Even within the same age group, children need to learn different things. A child who is scared of dogs needs to learn confidence as well as basic manners around dogs. A child who is confident with dogs needs to learn to respect dogs with which they are not familiar and to read their body language, as well as ways of mixing with dogs which are safe and enjoyable. Being scared and being over confident are opposite ends of the scale, but children at either extreme are at an increased risk of being bitten because of the way they behave.

The rules I teach include:

- Not all dogs are like yours at home. However they behave with their own dog at home (or a relative's dog, etc.), children must learn how to relate to unfamiliar dogs.
- Do not stroke a dog unless you know the dog's name. This is a useful guide for young children.

As the children get older, I recommend:

- First, ask the owner if you can stroke their dog, or give him a titbit etc.
- If the owner says it is fine, you should then 'ask the dog' by standing a little way from him, speaking to him and holding out your hand.
- If the dog approaches in a friendly way, go ahead and gently say hello, preferably by stroking him on the chest or shoulders.
- Never put your face close to strange dogs.
- Do not make excited gestures or noises.
- Never attempt to make friends with a dog which does

not have an owner with it. This includes dogs tied up outside shops as well as dogs running loose. A dog may look just like a friend's dog, you may think you know it, but still ignore it until an owner is with him.

If a child is unsure of a dog they should be taught to stand still, not scream or wave arms around, just to 'stand like a soldier'. It is best not to stare at the dog, but this is hard to do, so I often teach children that after standing still it is a good idea to look around slowly to find an adult. They should call to them and when they walk away, they should move slowly. If the child is holding a toy or food, I recommend that they drop it and back slowly away.

A Note about Health Issues

Health issues are always mentioned when discussing children and pets. It is beyond scope of this chapter to cover it. If you have any concerns about sick animals, consult your vet. I would advise a visit to the veterinary surgery as a chat with a veterinary surgeon will put any health worries into perspective. Appropriate treatments, vaccinations, fleas and worming at regular intervals, plus regular check-ups will keep most problems at bay. Teach children basic hygiene rules, such as washing hands after touching pets and before eating any food, and not to put faces near to pets. Take care with faeces, poop-scoop it and dispose of it effectively.

Conclusion

The subject of pets and children is huge and one which could be written about in far greater depth. The aim should be to avoid problems where possible by careful thought, socialising and avoiding difficult situations.

Where a problem exists or where there is a potential problem, a veterinary surgeon will be able to help by providing a referral to a recognised behaviour counsellor, such as a member of the APBC. Growing up with a pet for companionship should be a wonderful experience, with many benefits. Through careful thought and consideration, good points will be maximised and any possible negative points minimised.

One Final Note…

Whilst caution and common sense must always be exercised, most problems can be resolved or at least improved. Unfortunately, all too often, the general advice is that if your dog shows any aggression towards children, then you should part with the dog. This is usually based on concern for the child and a lack of knowledge on the part of the person giving the advice. Although great care must be taken, many problems involving aggression towards children can be overcome by following the right recommendations, and therefore avoid making hasty decisions that cannot be reversed.

Children and Cats

Donna Brander

The first thing to understand about getting any pet for a child is that all situations involving young children and pets are potentially dangerous. Interactions between pets and young children should always be supervised and every effort should be made to avoid any risk to both parties. Although the incidence of dog bites in the population has been well documented, injuries inflicted on children by the domestic cat are not often reported.

Nevertheless, being attacked by a cat can be both physically painful and profoundly disturbing emotionally to most children.

Some children are excellent with cats but kittens are fragile and easily injured. Kittens need to be handled properly in the early weeks of their life or they can develop some difficult behavioural problems. Cats and kittens are widely available and inexpensive and it is very easy to get either on an impulse; they may appear to be the easy choice for the family as they are perceived to be more independent and less of a commitment for the child or the family. This is partially true but can be part of the problem. Cats and children both want life on their terms and this can set up a conflict between them.

It is important to understand that cats can be very subtle in their signalling. Initial warnings can include mild skin rippling, flattening of the ears and turning away and ignoring overtures from people. Children do not always understand a cat's warning signals and this can lead to defensive aggression. A mildly irritated or anxious cat may lie with its paws on the floor (rather than tucked). It may flick its tail as a sign of irritation. This particular behaviour should not be mistaken as friendly tail wagging. Children should be taught the cat is saying that it wishes to be left alone.

The signals can become more overt, such as hissing, hair erection, and arching of the back, and if these more overt signals are ignored, the response can be swift. Most cats will attempt to move away at this point but some will resort to defensive aggression, particularly if confined or restrained. If the initial, subtle signalling is repeatedly ignored in interactions, the cat may rapidly resort to the more overt signalling. A child often becomes a victim of this behaviour because of a lack of knowledge of what the cat is communicating and how to respond.

Cats can remain reactive long after the triggering event. If a cat has had an unnerving meeting with another cat, dog or human, or even been frightened by noises, it may still be upset and aggressively responsive for some time. If a child attempts to handle the cat while it is still reactive, the cat may respond with what appears to be unprovoked aggression. It is best for the child always to initiate contact or play using a toy rather than touching it by hand. If the cat is overly reactive, it gives it the chance to attack the toy, or run away, rather than frightening the child.

Some cats may respond to unwanted handling and interaction by running away and hiding. This may solve the problem of aggression, but is unlikely to satisfy the point of getting the animal as a pet in the first place. Animals should be brought into the home to teach children to respect and understand them. Having a cat which runs away and hides, avoiding all interaction, hardly satisfies these criteria.

A fundamental consideration is whether or not a cat is an appropriate companion for your particular child. Choices are frequently made according to the parent's own desires and far too often the child is then made responsible for the animal. Consideration should be given to the temperament and capabilities of children and the demands they may make on the animal. If the animal chosen doesn't fit their demands and temperament, children may lose interest or, even worse, may grow impatient with the cat. The family should not have unrealistic expectations of the tolerance levels of a cat. All cats are individuals and have different tolerance levels and most cats can be defensively aggressive if they feel that no other choice is available.

When choosing a kitten for a child, look for a kitten that is responsive, energetic and curious. Shy kittens usually grow up to be shy cats and are not suitable for children. Older cats may be a better choice for children under seven or eight years of age. Check to see if the older cat has a calm response to noise, activity, and friendly overtures from children. Give the older cat time to settle into the home and allow him to make his own friendly overtures rather than having affection forced upon him. Give the cat control by teaching the children to leave the cat alone unless he comes to them. Point out friendly behaviour from a cat, e.g., head rubbing or tail in air, that means it is ready to be petted. Teach by example how cats like to be stroked (gently in the direction the hair lies) and how they like to have their cheeks rubbed.

You should make sure the cat has gentle playtimes and also be aware of inadvertently teaching a kitten inappropriate behaviour. Bring up your kitten as you mean for it to go on. It is very cute for a small kitten to attack your ankles but not so cute in a grown cat. Hands and feet should never be used as playthings and rough play should be discouraged between children and cats. Have a variety of toys available for the child and the kitten or cat to interact with.

Be aware! Cats require a lot of rest, in fact, most domestic cats sleep between fourteen and eighteen hours a day. Kittens have short, intensive play/eating periods with long periods of sleep. Some kittens have suffered malnutrition and even death because they were too tired to eat properly after they were handled and played with for too long. Kittens and cats need frequent rest periods all day long with food and water readily available when they are awake.

In some cases it may help to facilitate your cat's introduction to children by linking highly rewarding food with their presence as well as playing with toys. Dogs are rather omnivorous and soon link children with the bits of food they regularly drop. If you want your cat to link food rewards with children, make sure that it is food of high value to the cat, such as bits of special chicken-meat, so that the cat only associates it with being with the children.

Give the cat some control over its environment by providing boltholes that are not accessible to the child. These can be high places with box beds and blankets and food and water out of reach of the children. The addition of hiding places under a bed or in a cupboard can be beneficial. These resting places should be off limits to the children. This is to allow the cat the space and privacy it needs and teaches the children to set limits to their behaviour in order to respect the rights of others.

As with dogs, introducing a cat to an infant can begin long before the baby arrives. Try to get a tape of an infant baby crying and, perhaps, a dirty nappy. Begin to desensitise your cat to these novel items. You can even get a doll that moves and cries and give your cat a lot of positive attention, play and food rewards for being curious but calm around the 'baby'.

Make sure litter trays are private places for the cat. Many cats which ultimately have problems eliminating outside the litter tray have been disturbed while they were toileting in the tray. One unsupervised child was known to pick up and scold the family kitten when he caught it toileting in the litter tray. Needless to say, the cat began toileting behind the sofa and in the closet – anywhere hidden rather than in the tray.

A cat or kitten added to the family should definitely be seen as a family project rather than another playmate for a child. Cats should never be brought into the home to be living toys for the pleasure of the children and a cat may not be the right choice for children who have difficulty controlling their impulses.

Young children should be taught to think of having a cat as a privilege to look forward to when they are 'old enough', rather than as a right of childhood. Having a cat in the family should be about learning about a cat's needs and behaviour, how it communicates, how the child can fit into the cat's world and how to communicate appropriately with it. It should be about respect for the cat and learning responsibility for the care and welfare of another living animal.

Children and Rabbits, Guinea Pigs and Small Rodents

Emma Magnus

It is very common for children to be given a pet rabbit, guinea pig, hamster or gerbil in exchange for the dog or cat they really crave. The feeling amongst most parents is that these animals represent an easier alternative, a pet that will involve less care and expense. In addition they are readily available at most pet shops and can therefore be purchased on impulse with very little forward planning.

Unfortunately for many children the relationship falls at the first hurdle, with the pet becoming aggressive or simply 'not doing anything'. With a gerbil or hamster living for at least two years and a pet rabbit living for anything from five to ten years the purchase of these animals, especially for a child who may lose interest, should be as a result of an informed decision. Most of all, as already mentoned, it is important that pets are not purchased as a live 'toy' for a child's entertainment. Unless they are committed to providing for its needs and maintaining its welfare, albeit with parental help and guidance, it is better that they do not have a pet at all.

The husbandry requirements for each species of smaller pet are unique and care must be taken to ensure that they are being provided with the correct diet and the appropriate environment. For example, guinea pigs are unable to produce their own Vitamin C and therefore require special diets. A gerbil will become very destructive in a shallow cage with no room to burrow, it will become much more active in an aquarium with thick bedding to burrow in. A rabbit requires lots of daily exercise and social contact and may benefit from living indoors where it will be able to run around.

Rabbits, guinea pigs, rats, mice, gerbils and some species of hamster are very sociable and in the right situations a good relationship can develop between the child and the animal, but these are all prey animals and therefore have a very different perspective of the owner-pet relationship when compared with the cat or the dog. The interaction between a human and a prey animal involves an enormous amount of trust that can very easily descend into distrust and fear.

The greatest issue affecting a prey animal's relationship with a child is handling. For each species of rabbit, guinea pig or rodent there are a variety of breeds, differing in size and husbandry needs. Some of the larger rabbits are too heavy for a child to pick up; some of the smaller breeds of hamster are too small for hands that may squeeze too tightly. Whichever breed is chosen, it is important to get the animal when it is young, from a breeder who has ensured that it has been handled daily since it was weaned. The breeder, veterinary surgeon or another experienced individual should advise on the correct method of handling for the species. An easy mistake made by many pet owners is to pick a rabbit up with hands either side of the ribcage, with the rear end dangling. This method is likely to make the rabbit, a prey species swooped upon from above, feel unsafe, which may lead to some kicking and struggling. This is not only dangerous for an animal with a very fragile spine but can also lead them to avoid being handled in the future. The rabbit has a very fragile spine that must be supported at all times. Failure to do so can lead to back injuries as the rabbit struggles and kicks with the back legs. There are several methods suggested for handling rabbits that depend on individual preference and the size or age of the rabbit.

With the rabbit facing you, the first method is to grasp the rabbit firmly with one hand around the shoulder area, placing the ears flat against the back whilst the other hand takes the full weight of the rabbit under the bottom and scoops the rabbit on to your chest or lap The second method is to place one hand between the rabbit's front legs whilst lifting with the other hand under the bottom. Rabbits should never be lifted by the scruff or by their ears.

Similarly smaller pets have to be handled in a manner that is appropriate for their physical characteristics and their ethology. Once parents have researched what these are for the species concerned, they must ensure that their children fully understand and behave accordingly.

An animal which has not been handled regularly, or has experienced inept handling, is likely to display fearful behaviour each time the owner tries to pick it up, or even tries to feed it. This fear behaviour may be exhibited as cowering and hiding but it may also become aggressive behaviour. For a child, this is hugely disappointing and frightening and will generally lead them to avoid all contact with the animal in the future. For the rabbit this means life at the end of the garden in a hutch, being ignored.

5
The Importance of Positive Reinforcement

Inga MacKellar

Introduction

We live in a technology-driven age where people's expectations are demanding and instant solutions and responses seem mandatory. We expect fast, instant food; become annoyed when stuck in a traffic jam; become irritated at the supermarket checkout by the person who takes for ever to count out their money; we get cross with the dog when it's 'dawdling along', sniffing at every lamppost on its walk. I also challenge any computer-user to deny that at some stage they have not become frustrated with it for being too slow. Countless consumer programmes on the television and radio encourage us to complain about poor service and goods and to be less tolerant generally. This demanding lifestyle will invariably have a toll on our pets' lives – be it that we have less time for them or that we have high expectations and want them to respond to our wishes instantly.

There are many types of pet, but the dog remains one of the most popular and the species most frequently referred for behaviour problems (APBC). If we are to have a mutually enjoyable co-existence with our dogs, it is important that we make time for them, make an effort to understand them and think about how we interact with them.

I have found that many owners often resort to punishing their dogs for actions that they perceived to be inappropriate or unwanted, or perhaps for not obeying a specific command. In many instances, this was a result of a lack of understanding of dog behaviour and frustration. Due to the fact that dogs are frequently treated as humans *(anthropomorphism)*, many owners use 'human' rearing and disciplining methods.

In a study of dog owners in the Netherlands, owners were asked how they would react if their dog displayed an inappropriate behaviour. The problem behaviours cited were categorised into eight situations: fearful, aggressive/dominant, 'mating', emotional, bored/lonely, eating, unruly and disobedient. It was found that in most of these situations owners demanded compliance and reacted by addressing the dog severely. Punishing the dog was specified as the first option to use when the dog misbehaves and many of the situations elicited feelings of irritation and anger in the owners together with disappointment and powerlessness (Ben-Michael, 1997).

Owners frequently tell me 'He knows that he's done wrong as he looks guilty' and often feel hurt and let down that their much loved and cosseted dog is 'knowingly' disobedient and defying them. On further investigation these owners have, invariably, punished their dog for what they regard as 'bad' behaviour. They would automatically smack a dog if it had urinated in the house, shout at it to stop jumping up or, thankfully more rarely, give it an electric shock from a special collar to stop it wandering out of the garden. One of the most common punishments is reprimanding the dog on returning to its owner after it has 'run away' on a walk. Repeated use of punishment has not resulted in any improvement in behaviour and, in some instances, has resulted in problem behaviour getting worse. In most of these cases, considerable improvements were made after implementing positive reinforcement coupled with *negative* punishment.

How Dogs Learn

In order to understand how a dog might respond in a given situation, you must have some knowledge of how it learns. In the 1920s a Russian scientist, Ivan Pavlov, found that he could make dogs salivate even though food was not always present. He achieved this by letting the dogs hear the sound of a metronome just before they were fed. Over time, the dogs would salivate to the sound of the metronome even if the food did not appear. He could keep this up indefinitely as long as he sometimes gave food after the sound (Pavlov, 1927). This discovery is the basis of 'classical conditioning' i.e. a stimulus (e.g. the metronome), that has no apparent importance in itself, becomes significant because it predicts that something will follow (e.g. the food) and triggers an involuntary action (e.g. salivation). Classical conditioning can be used to trigger a response over which an animal has no control, including the process involved in emotional states. For example, if a dog gets excited about going for a walk, but is never taken on a lead, then the lead will have no significance and will not trigger excitement if it is produced. However, once a dog has been taken for a walk on a lead a few times, the sight or sound of it will trigger excitement in the dog, even if it is not taken for a walk.

Dogs also learn through 'operant conditioning'. In this instance a dog will learn that if it performs an action, that action will have a particular result. Simply put, the dog learns that 'If I do this...this happens'. There are four possible outcomes: positive reinforcement, negative reinforcement, positive punishment and negative punishment. Where the outcome of the action is something pleasant or beneficial, the dog is more likely to repeat that behaviour – the outcome positively reinforces the action. Conversely, if the outcome is something unpleasant, the dog is less likely to perform the behaviour – the outcome is aversive – it could be said that the dog has experienced positive punishment. Negative punishment involves the withdrawal or absence of a previously rewarding experience when the action is performed, which will *decrease* the likelihood of the behaviour being performed again. Negative reinforcement involves the withdrawal or absence of an aversive experience when the action is performed, resulting in an *increase* in the likelihood of the behaviour being performed.

Ineffective Positive Punishment

On the face of it the solution seems simple – if your dog does something that you don't want it to do, make the outcome aversive and it will not do it again. But behaviour is not that simple. What was the dog's motivation for the behaviour in the first place?

If, for example, a dog is behaving aggressively towards a neighbour because it is frightened of him – reacting in an aversive manner, such as the owner shouting, hitting the dog or pulling on its collar, may make the situation worse as this may further increase the dog's arousal. In some instances an owner's response can also act as a reward for the dog by giving it attention. A dog's fear may also be increased because something 'nasty' happens to it when the neighbour is present (being punished)! The next time the dog sees the neighbour it may be more frightened and react in a more aggressive way to try and scare off the neighbour... it gets reprimanded again... its fear further increases... and so the spiral continues.

Unfortunately, all the normal and usual human reactions to aggressive dogs tend to increase their aggression. In situations where a dog shows signs of aggression, trying to calm and comfort it reinforces the aggressive reaction and fear if present. Punishment increases the existing fear and thus leads to more aggression (Jones-Baade, 2001). However, by employing positive reinforcement methods the dog can be taught that it is rewarded when it does not react to the neighbour and, over time, it will start to associate the presence of the neighbour with something pleasant happening (a process called desensitisation and counter-conditioning).

Let's look at the common scenario of the dog which has ignored the recall command on a walk – when it eventually returns to the owner it is severely reprimanded. In essence, the dog has just been positively punished for returning to its owner. Dogs make an instant association with a particular action. Therefore, the dog does not understand that it is being punished for not responding to a command it heard ten minutes ago – it is just aware that it has approached its owner after having a nice time running around, sniffing in the forest, and the owner is immediately 'aggressive' towards it. Because approaching the owner has been aversive – the dog will then be more wary of returning to the owner next time... which makes the owner more frustrated and cross and results in the dog being told off more vehemently. A stage is eventually reached where the dog is reluctant to come back to the owner at all, so the owner resorts to walking the dog on the lead. If, rather than reacting with positive punishment, the owner had positively reinforced the dog with, say, a food reward when it had eventually returned, the dog would have learned that coming back to its owner meant something pleasant and would, in future, come back more eagerly when called.

Some owners believe that a dog should be instantly obedient because it respects them and that giving it rewards is bribery. However, bribery is very different from positive reinforcement. Bribery entails the use of a lure to encourage a particular action – positive reinforcement entails the giving of a reward after the action has been performed – a subtle but important difference.

Effective Positive Reinforcement

Most people immediately connect positive reinforcement, or reward training, with food. However, dogs are motivated by many things and what is rewarding depends on the drives, needs and temperament of the dog as well as the situation at the time. Obviously, where a dog is highly motivated by food, it makes sense to use food rewards. But if the dog has just eaten its dinner, it will probably be less motivated by food, and a toy or verbal praise may be more appropriate. Owners are sometimes concerned that their dog will become overweight if they utilise food rewards. This need not be so – a simple way to overcome this is to measure out the dog's food for the day and use part of the daily ration for positive reinforcement. Many dogs are now fed on a complete dry diet which can easily be carried in pockets

– ready to reinforce the dog positively. It is, however, important to have a variety of treats so that if the dog behaves particularly well, a more highly prized reward can be given. That is, there should be scales of reward. The better the performance of the required behaviour, the more appetitive the rewards should be. These can vary from pieces of chicken, ham and cheese to raw carrots and home-made liver cake. Touch, eye contact and reacting to the dog, as well as toys, etc., can also be added to reinforce a particular action positively.

The breed of dog will play a part in its motivation and it is important for owners to know the origins of the breed, as this will affect the dog's behaviour and general character. For example terriers were initially used (and still are in rural communities), for hunting. Most are extremely reactive and like squeaky toys, which they delight in shaking. Border Collies, bred for sheep herding, generally enjoy chasing and running whilst Cavalier King Charles Spaniels, which originated as companion dogs, particularly enjoy physical contact. There are numerous breed societies available to the dog owner and the greater the knowledge of the breed the easier it will be for the owner to find the right motivation. However, there will always be exceptions to the rule. For example, whilst it is generally accepted that Labrador Retrievers are highly motivated by food – there are some which are not. It is vital that owners are aware of what motivates their dog in order to use the appropriate method of positive reinforcement.

An important factor to consider with positive reinforcement is the type of reinforcement schedule used, i.e., how often the dog is rewarded. Whilst Gross (1992) defined five types of reinforcement schedule (continuous reinforcement, fixed interval, variable interval, fixed ratio and variable ratio), the two most commonly used in training dogs are continuous reinforcement and variable ratio:

- *Continuous reinforcement* – where each desired response is reinforced, e.g., every time the dog sits, it gets a piece of kibble.
- *Variable Ratio* – reinforcement is given spasmodically, so the reward is unpredictable, e.g., the dog is rewarded for sitting once, but the next time it is rewarded after sitting three times, then possibly twice, etc. – there is no set pattern as to when the reward is given.

As it is important that the dog learns what is required of it, continuous reinforcement is used initially to teach the dog a particular command or action. Once the owner is confident that the dog has learned what is required and it is responding each time, the rewards can then be given on a variable ratio basis. Variable ratio can be likened to gambling – the dog responds extremely well in the hope that it will be rewarded with a 'win'. If variable ratio coupled with a variance in the value of the reward, (i.e. sometimes the dog receives a piece of kibble and sometimes it receives a 'jackpot' – a handful of chicken), the dog's response will become even more reliable. This is because, sometimes, the dog is being positively reinforced and other times it is being negatively punished (it is not receiving it's expected reward) which leads to the dog becoming frustrated, trying harder and subsequently being positively rewarded for the increase in its behaviour. (This frustration effect is covered in more detail later in the chapter).

An extremely effective method of positive reinforcement is through the use of clicker-training, whereby the dog forms an association of reward with the sound of the clicker through classical conditioning. Once the

Photo: David Appleby

1. The Golden Retriever – selective breeding has enhanced its retrieving instinct

Photo: David Appleby

2. Exposure to a wide range of stimuli in early life enables the puppy to develop its maintenance set

Photo: Danny Archer

3. Puppies must learn good manners, when greeting each other on or off the lead

Photo: Danny Archer

4. Playing games with a dog is a great way of increasing the bond and helping children and dogs understand each other

Photo: Danny Archer

5. Family walks provide an opportunity for dogs to form positive associations with children

Photo: David Appleby

6. It is important to reinforce rather than ignore convenient behaviour

Photo: David Appleby

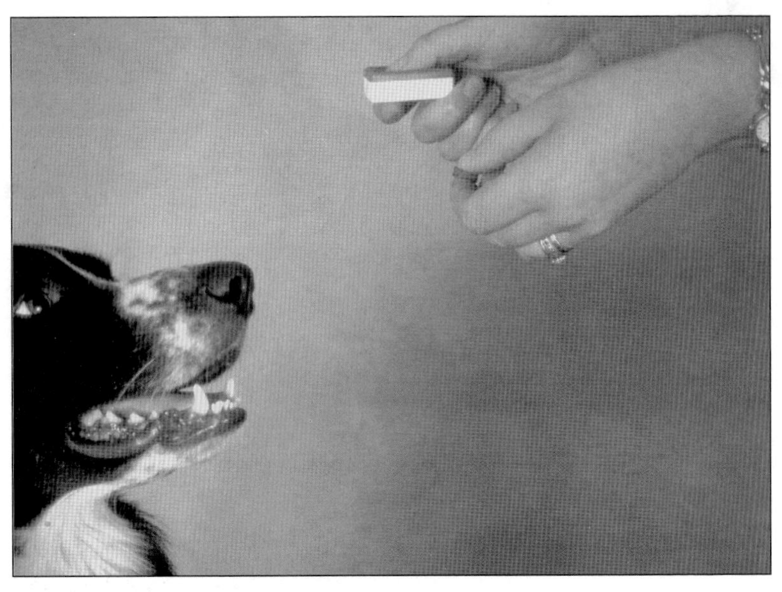

7. Clicker training provides an effective means of reinforcing the required behaviour

Photo: David Appleby

8. Scent communication allows signalling in the absence of the signaller

association is formed, this becomes a highly enjoyable and stimulating method of training for the dog as, in essence, it initially has to figure out for itself what it is being rewarded for. With a clicker it is easy to 'click' at the precise moment that the required behaviour is performed. Verbal or other signals can then be added as the dog starts to perform the action and it will make an association between the two. Eventually the dog will learn that performing the action when the signal is given will make the click treat occur, but they do not occur if the click treat is not given. Clicker training also helps owners to perfect their timing because they have to be vigilant and watch their dog in order to 'click' when it performs the required behaviour. There are a number of publications now available on this effective and fun method of positive reinforcement/negative punishment training.

Inadvertent Positive Reinforcement

Clients often tell me that they don't have time to be rewarding their dog constantly. However, many owners are masters of positive reinforcement – but don't realise it. Imagine the following scenarios with the family dog, Ben:

- Ben wants to play, but his owners are busy getting ready to go out for the night and take no notice of him. Ben happens to spot some training shoes lying on the ground, picks one up and starts to throw it in the air playfully. One of his owners instantly reacts and starts to chase him. Ben finds that this is great fun and runs around the house 'ducking and diving' with the owner chasing him in a frantic manner. The owner chases Ben because he is desperate to get his expensive shoe back. Ben views this interaction as a game and through repetition he could learn that it is a good way of getting attention. The owner positively reinforced the dog's behaviour of running away with the training shoe.

- An owner is sitting watching TV and Ben comes approaches and lays his head on their lap. The owner absentmindedly strokes Ben's head. Eventually, the owner stops stroking his head and Ben nudges the owner with his nose and they immediately respond by stroking him again. The owner has positively reinforced the dog's attention-seeking behaviour.
- The owners are eating dinner and Ben approaches the table. One of the owners glances at and makes eye contact with him, which encourages Ben to sit and look 'lovingly' at them. Ben continues to try and catch his owner's attention by placing his head on the owner's knee (which he knows has worked in the past) at which point, another member of the family may comment, 'Don't be mean – give Ben a bit of your steak' to which the owner responds 'I'm not supposed to feed dogs from the table . . . am I Ben?', looks at him and slips him a bit of his steak. Ben has been positively reinforced for begging at the table.
- The owner returns from work, walks into the house and Ben rushes up and starts to jump up at him in excitement. The owner is delighted by this effusive show of affection and interacts with Ben by greeting and patting him. The owner has positively reinforced Ben's jumping behaviour.

All these scenarios have involved inadvertent positive reinforcement on the part of the owner and made Ben more likely to behave in a particular manner when the situation and similar situations are repeated. Yet, more often than not, when Ben has been lying calmly or chewing a bone (rather than his owner's training shoes), he has been totally ignored. Many owners tend to interact with their dog when it is, from their perspective, 'misbehaving' or instigating interaction, but ignores it when it is behaving well.

None of the types of behaviour mentioned in the previous scenarios is necessarily a problem and many owners happily accept them. However, imagine a scenario where a young girl comes into the house and Ben bounces up to greet her, as he has done so for many years, and knocks her over. The girl cries, the owners are mortified and immediately punish Ben by shouting 'Bad Boy!' and giving him a smack. Ben is totally confused because he has behaved in his normal way, for which he has always been rewarded, but he is now being told off. Ben is unsure of the situation so he resorts to behaving in a manner that he has previously been rewarded for – he jumps up more vigorously. The owners, who are still attending to the crying child get cross with Ben and banish him to the kitchen. When Ben is eventually let out, he approaches his owners, cowering, tail between his legs and ears flattened. The owners then 'forgive him' as they believe he is looking 'guilty' because, to them, he looks as if 'he knows that he's done wrong'. Ben does not know that he has done 'wrong' – he has been stressed by his owners' behaviour, he anticipates more owner aggression and the resultant emotional state of fear is manifested in his body language. This situation could have been avoided if Ben had been accustomed to being positively reinforced for acting calmly when the family came home rather than for leaping about.

The Dangers of Positive Punishment

So why not use positive punishment? Surely the dog needs to know 'who is boss'? There are an ever increasing number of gadgets and 'quick fix' solutions available for owners with problem dogs and a considerable number of these involve the use of positive punishment, some confrontational (directly involving the owner) and some remote (the dog does not make a connection of punishment with the owner). These range from check chains to collars that automatically squirt citronella spray when a dog barks to collars that deliver an electric shock. All these methods can be misused either accidentally, or intentionally, and can result in pain or other distress for the dog.

Another possible repercussion of aversive devices can be the development of fearful behaviour towards an unrelated object or environment. For example, if a shock collar is activated to stop a dog barking and at that same instance a child walks past, the dog may associate pain with the presence of that child and start to display fear-aggression towards them next time it sees them. I have seen a case where an owner had installed a special fence to stop her boisterous Collie wandering off (This device works by giving the dog a shock through a special collar when it gets close to a boundary). During the consultation it emerged that when the other dog in the house, a timid Collie, had the special collar placed on him he wouldn't even go into the garden. One could argue that the device had worked, but to such an extent that one of the dogs was now frightened of going outside at all.

Whilst accurately executed positive punishment, implemented by someone who is aware of all the possible dangers, can be effective in stopping unwanted behaviour, most people do not have the expertise to implement it without potentially detrimental effects. The timing of positive punishment is absolutely crucial and a delay in its use can have profound implications for the effectiveness of the technique (Solomon, Turner and Lessac, 1968) and the welfare of the dog. The average dog owner rarely 'punishes' their dog immediately the behaviour is displayed – a prime example being that of the dog which urinates in the house when it is left alone and is then punished some hours later when the owner returns. The dog will not make any association between the punishment and its previous urination, but it may start to make an association between the owner's aggression and their return home. The problem may well escalate as the dog displays increasingly stressed behaviour manifested in its body language when the owners return. If it finds that this behaviour stops or prevents their aggression it could learn to use it to avert its owner's 'aggressive' behaviour towards it on future occasions. If a dog is extremely stressed, a possible consequence is urination and the dog may start to urinate as soon as it sees the owner – which would exacerbate the situation further.

Another point to consider is how severe should the punishment be? A verbal reprimand? The use of aversive tasting substances such as hot mustard or chilli sauce smeared on items to stop a dog chewing? A smack with a rolled-up newspaper? Hitting it with a stick...or worse? Positive punishment is intended to be aversive, but it is clearly undesirable to inflict more pain than absolutely necessary. But what is necessary?

If the punishment is too mild, it may be ineffective, and the problem may then not be remedied easily by switching to a stronger punishment. In a study, pigeons were punished with an electric shock for pecking at a key that produced food (Azrin, Holz & Hake, 1963). When the shock was set at 80 volts the behaviour was totally suppressed. When the voltage was set at 60 volts, however, it had little effect, and when its intensity was then increased in gradual steps, the pigeons still continued to peck at the key for food even when the voltage had reached 300 volts. If punishment intensity is set at low levels initially and then increased gradually it may prove ineffective because the dog adapts to the gradual increasing intensity. The optimum level of punishment is thus the minimum level that is effective in suppressing the behaviour – but choosing that level is a skill that few people can claim to have. However, even if they have it, is it necessary or desirable? What are the implications for the dog's welfare and its relationship with its owners?

Cognitive and linguistic capacities have been found to play a role in determining human reactions to punishment, but dogs do not have linguistic capacities. A study by Cheyne (1969) found that in a group of children who were given a number of toys and then punished for playing with a particular one, those children who had an explanation given to them as to *why* they mustn't play with the 'forbidden' toy were far less likely to play with it in a future situation when compared to those children who were just told 'No' and not given an explanation. Sears, Maccoby and Levin (1957) found that mothers who made extensive use of reasoning, reported punishment to be far more effective than those who used punishment alone. Whilst many of us tend to speak to our dogs on a daily basis, and they will respond to certain words such as 'Sit' or 'Down', they do not understand what we are saying. Therefore, in a situation where a dog is being reprimanded, a reasoned explanation of why they are being punished will have no effect whatsoever, and the dog will just become confused and may develop emotional responses related to stress and anxiety in other aspects of its relationship. In extreme cases, a dog which feels particularly threatened may resort to defensive aggression towards its owner in an attempt to stop the physical punishment and in anticipation of it. This can then result in an extreme breakdown of the owner–pet relationship, and in some cases, the euthanasia of the dog.

It is important to remember that, as detailed previously, if a dog is displaying fear-aggression, positive punishment can lead to an increase in that fear and an increase in the intensity of the behaviour.

If a dog is being positively-punished consistently, it can also result in an increase in stress. Animals (and humans), placed in a situation, where they have to deal with a threatening situation will experience a number of physiological changes, one of which is the increased production of adrenal gland hormones. This physiological response is useful as a short-term response in an aversive situation as it results in increased blood flow to the muscles and provides quick mobilisation of energy for vigorous movement, but it is harmful in the long term. Prolonged increases in the level of glucocorticoids (cortisol) can damage muscle tissue, inhibit growth, inhibit the inflammatory response and suppress the immune system (Carlson, 1994). Animal research suggests that stress can encourage the growth of malignant tumours and exacerbate allergies and auto-immune diseases. Equally, prolonged stress can impair the learning process (Dweck & Licht, 1980). Therefore, if a dog is continually placed in a situation where it is being positively punished by its owners, its physical as well as its cognitive well-being is compromised.

The Action of Negative Punishment

So far I have talked about positive reinforcement and positive punishment. There are, however, two other learning scenarios to consider – that of negative reinforcement and negative punishment. Liebermann (1993) defined negative reinforcement as an increase in the probability of a response due to the removal of an aversive stimulus, and negative punishment as a decrease in the likelihood of a response due to the removal of a reinforcing stimulus .

To illustrate these two scenarios, let us consider the situation of two children playing – a boy and a girl. The girl is very spoilt and used to getting her own way. She keeps on pulling the boy's hair, telling him that she will not let go until the boy gives her a sweet. Here we have a situation where negative reinforcement is being utilised – the pain (positive punishment) will continue until the boy reacts in the required manner – relinquishing a sweet. This particular method of learning, negative reinforcement, is, obviously, not one to be recommended, as its welfare implications are related to positive punishment.

On another occasion the girl continually pesters the boy for a sweet because this has always worked with her parents. Her pleas, cries and tantrums are incessant, but the boy takes no notice, she becomes increasingly frustrated and her behaviour escalates, she stamps, cries and throws herself on the ground but all to no avail and the boy continues to ignore her. Eventually, exhausted, she stops her tantrum and sits down calmly, at which point the boy gives her a sweet. Although her tantrums had previously worked, in this instance, she will have started to learn that tantrums will not result in obtaining a sweet (negative punishment), but sitting calmly will (positive reinforcement).

Negative punishment involves what is termed 'the frustration effect' where the subject – be it a person or a dog – when faced with a situation where he/she has previously been rewarded for their behaviour, will get increasingly frustrated and will try harder but can become angry or depressed if the expected reward does not materialise. There appears to be a strong emotional component – think of the angry man kicking a vending machine when his expected can of Coke doesn't materialise despite frantically pressing all the buttons, or the frustration that you feel when you turn the car ignition on and the engine is dead. You repeatedly turn the key and possibly start shouting at the car before you accept that the engine will not start. Obviously in these two examples, in reality, the behaviour would not be extinguished as the vending machine will eventually be replenished with cans of Coke and the garage will fix your car so that when you turn the key the engine will start again. However, if, for example, the vending machine was never replenished again, the man would eventually learn that kicking or repeatedly putting money in that particular vending machine was ineffective. As he would not receive the 'reward' of the Coke can, he would eventually stop trying – his behaviour would have been extinguished.

Usefulness of Negative Punishment

Unlike positive punishment (where an action results in an aversive outcome), negative punishment entails the absence or withdrawal of an expected reward. This method can be used to stop unwanted behaviour that has previously been either consciously or inadvertently positively reinforced. Negative punishment can be a difficult concept to grasp, I believe, because the terminology is confusing. The words negative and punishment both conjure up images of something aversive, whereas the concept could be described more simply as non-reward or the ignoring of a behaviour that was previously responded to.

To illustrate this, let us return to Ben, who instantly leaps up at everyone who enters the house. How can we stop such behaviour?

In many instances owners will resort to waving their arms about and shouting at the dog to 'Get down' or, possibly, hitting it. The first will inevitably prove to be ineffective, because Ben is stimulated by the increased interaction (positive reinforcement) and will continue to jump up. Hitting the dog will confuse him and he will redouble his efforts to display the behaviour that he has learned is rewarded by his owner – jumping up. This scenario continues on a daily basis with the owners becoming ever more frustrated and the dog increasingly confused. So what can be done?

Employing negative punishment will initially result in frustration but eventually lead to extinction of the behaviour. All people entering the house must completely ignore Ben when he leaps up – including no eye contact, ie. they need to behave as if Ben doesn't exist. Initially, Ben will be frustrated as he will not receive the response that he expects and will increase his efforts i.e. the behaviour will get worse. (It is at this stage when I am explaining the procedure to clients that they interject 'But we've tried ignoring him and it doesn't work – he just gets worse!' When I ask how long they persevered with the 'ignoring regime', they will invariably tell me that the dog became such a 'pain' that they couldn't ignore him any more and they gave up.) It is vital that the owners do not succumb to interacting with him, as otherwise Ben will learn that if he leaps up for long enough and hard enough, he will eventually get a response. However, by completely ignoring him every time he is excited no matter how long his excitable behaviour persists, and positively reinforcing the calm behaviour by greeting him, Ben's excitable behaviour will eventually be extinguished and calm behaviour, which is differentially reinforced, will be more likely to occur.

The problem behaviour will have been solved through the use of negative punishment coupled with positive reinforcement, without the use of any positive punishment and without damaging the relationship between dog and owner. Obviously, if Ben had been positively reinforced for sitting calmly when family or visitors arrived at the house from the outset, the problem behaviour would never have developed in the first place.

Negative punishment techniques can also be combined with 'conditioned avoidance', where the dog learns that it will fail in a particular action if a certain stimulus is present. Conditioned avoidance can involve the use of training discs (five small cymbals on a key ring). Once the discs have been properly introduced, the dog will associate failure (non reward) with the sound of the discs and will stop whatever behaviour it is displaying when it hears the sound. For example, a dog jumping up to get food from a kitchen work surface can be made to jump down and ignore the food by the owner chinking the discs or throwing them to the ground. Training discs can be used to stop all manner of behaviour from some types of aggression to jumping up.

Difficult Situations

Positive reinforcement (coupled with negative punishment) can be successfully used in most situations to overcome problem behaviours. This can sometimes be difficult to imagine when you have a dog on the end of a lead pulling its owner over in an effort to bite the next-door neighbour.

Aggression is one of the most common behaviour problems in dogs and one to which owners frequently react in an equally aggressive manner, positively punishing their dog. Quite often, if the aggression is directed towards another dog or person outside the home, the punishment given to the dog will be as a result of embarrassment, and the owner feels that they have to be seen to tell their dog off for its behaviour. I always advise owners, difficult though it may seem, to concentrate on reacting to their dog in the correct manner, rather than fulfilling another person's expectations. If a dog has behaved unsociably it is always important to apologise calmly and politely, but then concentrate on the dog and try to think why it behaved as it did.

If their dog displays aggressive behaviour many owners are reluctant to muzzle it as a precaution. I would suggest, once again, this is primarily because they are embarrassed or worried about what other people think or because it does not look very nice. However – if an aggression problem exists, it is far better to ensure everyone's safety by placing a basket muzzle on the dog. This will also enable the owner to be more relaxed because he or she will know that, should their dog behave aggressively, it will not be able to cause any serious injury. It is important, however, to introduce a dog to wearing a muzzle gradually rather than forcing one on to it in distressing situations.

Many potentially serious situations, such as sheep chasing, can be easily avoided. If an owner knows that this is problem behaviour, the simplest solution is to make sure that the dog is walked well away from any sheep, and kept on a lead when necessary.

As a general rule I recommend the following solutions, depending on the scenarios involved.

- If the unwanted behaviour is not serious, but something that the owner wishes to modify, such as incessant attention-seeking (barking, leaping up etc.), ignore the behaviour (negative punishment) and follow up with positive reinforcement of the desired behaviour. However, if the dog has an emotional need, you may have to provide an outlet for its attention-seeking behaviour, by using an interactive key, for example – where the dog learns that picking up a specific object, such as a particular toy, will result in attention from the owner.
- If behaviour needs to be stopped immediately, try interrupting, distracting and positive reinforcement. For example, if a dog is about to chase a cat (a very rewarding experience for the dog), a distinctive sound, such as a squeaky toy, can be used to obtain the dog's attention, before it starts to chase. If this is immediately followed by encouraging the dog, in an excited manner, to come to the owner and be positively reinforced with the squeaky toy and subsequent game, this will result in the dog concentrating on its owner rather than

chasing the cat. Eventually, the dog will find it more rewarding to ignore the cat, as it will learn that it receives stimulating interaction from its owner.

- Some dogs respond more readily to the use of a 'leave' command. It must be reintroduced in increasingly stimulating circumstances so that the dog learns that it is positively reinforced for its appropriate reaction to the signal of 'leave'. For example, the owner can progress from calling their dog away from a stationary toy, to one that is moving slightly, to calling it away from one moving at speed. Eventually, when the dog is put in a situation where it may have previously chased a cat, for example, it will return to its owner once the signal 'leave' is given and, due to the implementation of positive reinforcement, its motivation to chase the cat will be reduced.
- Where a behaviour must be stopped so that positive reinforcement of an appropriate behaviour can occur, the use of 'conditioned avoidance techniques', such as training discs, can be extremely effective, as they can help owners deal with more simpler behaviour problems. However, it is important that owners understand the concept properly before using them. Children must be prevented from playing with the discs, as indeed they must be prevented from playing with a clicker, as this will confuse the dog and make solution ineffective.

Problem behaviours, particularly more serious ones, can be very complex and involve a number of contributory factors such as the history of the dog, the relationship with the owner, how long the behaviour has been displayed, how much stimulation the dog receives and so forth. It is vital to assess the motivation behind the behaviour accurately. A solution may be to recommend a number of methods of behaviour modification, such as the use of conditioned avoidance, an interactive key, house lines, etc. It is not possible to detail methods for specific problems in this short chapter and those that have been included here are only examples. Owners who are worried about their dog's behaviour should seek professional behavioural help from an APBC member so that the problem can be understood properly, and an appropriate behaviour modification plan recommended and properly implemented.

Summary

I will not deny that aversive methods can sometimes be effective in stopping a dog's inappropriate behaviour. But in most instances where positive punishment is implemented, it is mistimed and without awareness of the possible consequences. These may vary from increasing the unwanted behaviour, such as in fear aggression, to the dog forming an association of pain with an unrelated event, such as a child passing by. Frequent positive punishment also leads to stress, both for the owner and the dog, and a breakdown of the owner–dog relationship. Surely it is better to have a dog which responds to signals because it has learned that compliance results in something pleasant, rather than a dog which responds because it is frightened and wants to avoid punishment?

Someone once said to me that they couldn't stand by if they came across a group of boys throwing stones at a window, they would have to 'sort them out'. Whilst they did not clarify what methods would be used 'to sort' the boys, I rather suspect they would have included some sort of confrontational positive punishment.

REINFORCEMENT AND PUNISHMENT	
POSITIVE REINFORCEMENT ✓	**POSITIVE PUNISHMENT?**
A rewarding outcome for a particular action: e.g. ◆ *When the dog sits calmly, it is stroked.* ◆ *When the dog is not attention seeking it receives attention* ***CONTINUOUS REINFORCEMENT LEADING TO A VARIABLE RATIO REINFORCEMENT SCHEDULE*** ***STRENGTHENS OWNER-DOG RELATIONSHIP***	An aversive outcome for a particular action: e.g. ◆ *The dog is jerked on the lead when it does not sit.* ***POSSIBLE ASSOCIATION OF PAIN WITH UNRELATED STIMULUS*** ***TIMING AND INTENSITY OF PUNISHMENT MUST BE ACCURATELY IMPLEMENTED*** ***WELFARE IMPLICATIONS*** ***CAN COMPROMISE OWNER-DOG RELATIONSHIP***
NEGATIVE REINFORCEMENT ✗	***NEGATIVE PUNISHMENT ✓***
The absence or withdrawal of an aversive experience when the dog displays the required behaviour e.g. ◆ *Jerking on lead stops when the dog sits.* ***COMPROMISES OWNER-DOG RELATIONSHIP WELFARE IMPLICATIONS***	The withdrawal of a rewarding outcome where the action performed was previously positively reinforced: e.g. ◆ *Ignoring the dog when it jumps up.* ◆ *Ignoring the dog when it seeks attention.* ***IMPLEMENTED ON A CONTINUOUS NON-REINFORCEMENT SCHEDULE***

My aim, and undoubtedly that, of my APBC colleagues, would be the same – to stop the boys – but our approach would be different. In the first instance, we would interrupt and distract them from the behaviour but then try to find out why they were behaving like this. Boredom? Frustration? They didn't like the person living there? Whatever the reason . . . once ascertained, the core motivation could be assessed and the problem behaviour dealt with to obtain a long-term resolution. Admittedly, instant positive punishment may well stop the boys' behaviour on that day – but it would not have dealt with their motivation and if the motivation to perform the behaviour is stronger than the fear of the potential punishment the behaviour will continue. However, once the motivation is known the behaviour can be redirected into an appropriate, more motivating, positively reinforced action, and the initial unwanted behaviour will stop.

The majority of owners have dogs as pets. They are considered to be a part of the family, trusted and faithful companions. However, all too often, we totally ignore them when they are behaving well, be it lying calmly, interacting well with people and other dogs, or not chasing the cat. But we instantly respond and interact with them when they start to behave inappropriately. Dogs do not speak the same language as we do – we cannot explain things to them; they will only learn from our interaction with them. So it is important that we, as a supposedly superior species, through patience, understanding and positive reinforcement, nurture, develop and continue to maintain that most cherished and potentially rewarding of relationships with man's best friend.

6
Scent Communication in Dogs – What We need to Know and Why

Charlotte Nevison and Rosie Barclay

Are you embarrassed when your dog thrusts its nose into the groin of a visitor to the household? Has your usually obedient canine ever been known to run off, apparently oblivious to your calls, then stop and roll delightedly in the remains of a long dead creature or fox scat? Do you have to wait because your dog has refused to move until he has checked out and cocked his leg at every lamp post, or seen your otherwise placid pooch turn into a master escape artist when the bitch next door is in heat? Yes? Then you are not alone. To us these behaviours can be inexplicable, confusing, inconvenient and possibly repulsive, but these behaviours indicate that our canine companions are motivated to respond to scents that we are either unaware of, or do not appreciate. In this chapter we will unravel some of the mysteries of the sensory world of dogs by illustrating how their sense of smell differs from ours, why scents are important to them and how misunderstandings about scent relate to some of the common 'problems' we experience with our canine companions.

I Sniff Therefore I Smell... The Olfactory Capabilities of Dogs

Before discussing what important information is conveyed in scents we need to think about what scents are and how animals gain information from them. You will be aware that objects may smell, as may liquids and that some smell more strongly than others. You may also become aware of an odour without deliberately sniffing at something. To this end all animals are basically alike but what is your brain actually responding to?

Scents are airborne molecules (volatiles) that have typically been released from a solid or liquid source. They may consist of a variety of volatiles released at different rates depending on factors such as the weather and whether or not they are being sniffed. When inhaled, volatiles are detected by specialised cells which send signals to the brain. These signals may then elicit a behavioural response *if* the volatiles convey information that is relevant to the animal at that particular time.

Although we detect odours in the same way, in comparison with most other mammals, humans are pretty poor at detecting scent. It is estimated that dogs have a 100–1000 times greater potential to detect odours than we do, because they possess more receptors for scent. This allows them to detect volatiles at a lower concentration in the atmosphere; they can more easily detect sources which release volatiles slowly or that are further away.

Making Sense of Scents – What Is the Relevance of Smell to Dogs?

Scientists have demonstrated that dogs and other mammals have a much superior sense of smell to us and are better detectors than any machines ('artificial noses') that have been developed to date. Such research is largely driven by the fact that detecting scents can be of considerable use in human society. Dogs are trained to detect drugs, firearms and distinguish human scents during search and rescue operations. However, though these skills are remarkable, they are learned responses acquired through the process of *training*. They reflect how amazing the canine sense of smell is but they do not tell us why it has evolved to be so special. Put simply, dogs are excellent scent detectors because this skill was of use to their wild ancestors.

Although dogs have been domesticated for centuries, scents that were naturally motivating and reinforcing to their ancestors are likely, still, to influence the untrained behaviour of the domestic dog, though responses may be tempered by breeding, training and whether or not their needs are fulfilled in the domestic environment.

What Scents are Important to Dogs, and Why?

Two of the key objectives for ancestral dogs was to locate food and defend physical resources as part of a cohesive social group, all within a complex and possibly large area. Scents played a key role in achieving these objectives.

SNIFFING FOR THEIR SUPPER

For a dog 15,000 years ago there were no easy meals and no bins containing tasty scraps. Potential prey would have been scattered and constantly moving over rough terrain, not neatly penned up in fields. In short, the chances of seeing potential meals from a distance were slim.

Hearing would have been of limited use because a dog's hearing extends to ultrasonic frequencies, which are easily obstructed by objects, including hilly terrain and trees, which may have been more important when dogs were evolving.

Fortunately for ancestral dogs (and other predators), potential prey species, or their waste products or scent marks, would have smelt as they released volatiles. Carcasses also smell. Such scents would travel through the atmosphere, around potential obstacles, and would become stronger as they were approached (thus indicating where the source was). Once the food was located the remaining problem to ancestral dogs was acquiring it. The ease at which this was achieved would depend on factors such as its abundance, size and the environment, but evidence suggests that much of the time they would have needed a little help from other pack members, which in turn required an efficient means of communicating with each other.

THE SOCIAL SIGNIFICANCE OF SCENT

Much of our understanding of ancestral canine behaviour comes from observations of closely related wolf species and modern-day feral dog packs. Pair bonds have been reported and are especially important when considering relationships such as the bitch-pup bond. However, free-living dogs typically associate into larger groups, packs, which consist of a number of individuals. Collectively animals within a pack defend a territory, although animals are not of an equivalent status.

Although members within a pack are socially bonded, territories are such that individuals within a pack are not constantly in each others' presence. So it is essential for group members, upon meeting, are able to recognise each other as part of their social group. They also need to communicate with each other remotely. Non pack members have to be identified, and where necessary excluded, as they may compete for vital resources or mates. Scents play a major role in achieving these objectives.

WHAT INFORMATION DO SCENTS CONVEY AND HOW?

Scientific understanding of scent-marking behaviour is largely drawn from studies of rodents, in particular mice, which are easier to study than large groups of dogs. From a combination of body scents and the distribution and properties of scent marks it appears that mice can determine an individual's identity, sex, relatedness, social status, health status and how long it has been since a scent-mark depositor was there. They can then use this information to determine how to respond to particular individuals if and when they meet and whether or not it is safe to stay in the area.

It is likely that dogs may also gain social, temporal and maybe even health information from scent marks which, with visual, tactile and auditory support, may determine their behaviour in both domestic and feral situations. Our understanding of the social relevance and behavioural effect of dog scents, though currently not as detailed as it is in mice, is growing. Recent advances, including behavioural and molecular studies of a pheromone secreted by bitches that calms her pups, has given us a more detailed understanding. By creating synthetic versions of these pheromones, such as dog appeasement pheromone, (DAP) we also gain potential tools which, with care and appropriate guidance, may be used to modify behaviour in a beneficial manner in a domestic environment.

In the following section we review what we do know about how dogs use scents to communicate and how we can use it to gain a better understanding of our domestic companions and their 'problems'.

Scents and their Sources

Upon meeting dogs use a combination of postures, sounds and body scents to gain information about each other. (Who is this? Is he a threat?)

The big advantage of scents over visual and auditory signals is that they can be deposited in the environment as marks which gradually release volatiles to convey social information about the depositor in his or her absence, much as we leave messages with a pencil and paper. Unlike the perception of sound and vision, scents are less disturbed by the environment and have a relatively long life.

Odours that provide socially important information in mammalian species are known to be released from many locations on the body including the mouth (saliva), urogenital tract (urine, vaginal fluid) or released from specialised secretory glands near the skin surface (oily or watery cellular secretions). Dogs may use all these substances to convey social communication, though to human observers the deposition of urine marks is the most apparent marking behaviour and has thus drawn the most attention.

URINE-MARKING

Urination by dogs serves two functions: it removes potentially harmful liquid waste from the body (elimination) and it conveys socially significant, individually specific information to other dogs (marking), typically in the absence of the marking animal.

Males mark in a very characteristic manner by depositing their urine in brief directed bursts by raising one of their hind legs; high if the urine is directed at a vertical target and lower if at a horizontal target. Urine may not be eliminated on up to a third of these occasions (often also observed in pet dogs), though figures vary widely between studies. Occasionally males may squat to urinate, particularly if castrated, which is the typical marking posture of females. Occasionally females may raise a leg into a more male-like posture.

In feral dog colonies, and indeed our pet dog population, adult male dogs urinate about three times as often as females. Most of the time males will expel their urine as marks, while females urinate mainly to relieve themselves, but may mark to advertise their reproductive status.

The positioning of scent marks can be particularly revealing when it comes to determining their purpose. Detailed studies have shown that marks are deposited mostly at courting sites. The sexual basis of urine marks is supported by studies, which have shown that males are more attracted to urine odours of females than vaginal or anal sac scents, particularly when females are in heat. In the presence of females, males will keenly and possessively overmark female urinations probably to convey ownership and status (though mating in feral groups is typically not exclusively between known breeding animals). However, overmarking (and queuing to do so) is also seen within sexes and is often seen in domestic dogs and wolves. It is not clear whether this serves as an indicator of status by the overmarker or to make a 'family' mark at significant points in the territory.

Marks are also frequently deposited at feeding sites and territory boundaries with nest sites being rarely marked. This suggests that the function of scent-marking is two-fold. The first is to attract mating partners (both sexes) and claim them (males). The second is to claim and defend territorial borders (which in dogs will often partially overlap with other packs) and feeding sites. Nesting sites are probably not marked with urine as to do this would draw attention to youngsters.

Though all individuals do appear to mark, there is variation in the number of times different individuals of the same sex mark. This is likely to be related to their position in the social hierarchy. Studies have shown that individuals who attain a high status in the pack (particularly males) generally mark at a higher rate, and further increase their marking frequency when their position in the hierarchy is challenged. It is quite possible that such changes in social behaviour are related to changes in the animal's emotional state. From a functional perspective, in natural environments depositing marks at territory borders takes time and energy and exposes individuals to greater levels of stress and danger. Therefore the ability to do this constantly is thought to reveal the individual's superior qualities and thus be attractive to the opposite sex, and deter or combat challenges (and thus the chance of injuries) from other males. In mice, individuals of a lower status will typically inhibit their marking response, depositing their urine less often and in larger amounts which appear to act as a message to higher-status individuals that they do not pose a challenge to them. Whilst it is unclear whether this happens in dogs, the release of large volumes of urine (and also faeces) is widely interpreted as being an indicator of anxiety in many species, and this is likely to correlate to an individual's social status.

FAECAL MARKING AND ANAL GLAND SECRETIONS

When a dog defecates a small amount of fluid may also be secreted from the anal glands. Rarely, anal glands may also be voluntarily discharged without defecation, but such behaviour is much more common in the more distantly related cousins of the dog, such as hyena. In wolves, deposition of faeces along territory borders and trails suggests that they may play a role in designating ownership of an area.

Both faeces and anal gland secretions have been considered as sources of socially significant scents in dogs relating to sexual attraction, territory marking or signalling of danger. But what is the evidence?

Anal sac secretions contain a solution of lipids, particulate material and bacteria which help to produce an odour. It is thought that the number of volatiles identified in anal sac secretions and the between sample variation is sufficient for unique combinations to occur in different dogs and could therefore be used as a label to convey ownership of territories. Although deposition patterns in wolves suggest this is a possibility in that species, there is virtually no evidence for this in feral or domestic dogs. There appears to be little attraction of potential mates to anal gland secretions, at least relative to urine marks. Anecdotal evidence suggests that discharge of anal sacs occurs with fear- or anxiety-inducing situations. This has not, however, been supported by examining the responses of dogs to these odours.

OTHER POTENTIALLY IMPORTANT SOURCES OF SIGNIFICANT SCENT

The sources of scents discussed above and their relevance to dog communication have drawn most attention from scientists interested in theories of animal behaviour. Other pheromones, such as that secreted by bitches which acts to pacify their puppies, have raised interest, not only because of what they reveal about canine behaviour per se, but also because of their potential use in an applied context in the domestic environment. However, dogs secrete many other substances which could also be important to them, some of which are described below, which have yet to be studied in detail.

Upon meeting, dogs are typically drawn to investigate the ano-genital region where they may be particularly concerned with information conveyed in urinary odours. They are also drawn to investigate the facial region though clearly this involves an element of danger. Licking may also occur, which in wolves is usually directed by the more inhibited individual of any pair. Domestic dogs are also often seen by their owners to lick other areas of each other's bodies (notably mouths and ears) and dogs will often also lick their owners, which owners may reward or discourage. Dog saliva is known to have some anti-bacterial properties and thus licking may have some health benefits and mutual grooming of each other's hard-to-reach areas may play a role in social bonding. However saliva, distributed by licking, may also have a communication role. Proteins found in dog's saliva have a notable degree of similarity to those that mice, and other species, use to communicate their identity. So it is plausible that saliva may be used to convey identity, ownership or group membership.

Domestication and Domestic Environments – Its Influence on Scent Communication Between Dogs

Given the sensory differences that exist between dogs and humans, it is remarkable that by and large we co-exist so amenably. Through the gradual process of association for mutual benefit that led to the domestication of dogs we have managed to learn to understand each other to some extent, but not entirely.

During the process of dog's domestication we have altered many factors that influence how dogs behave. Dogs rarely live with groups of other individuals of their species, typically living as the only canine or one of a pair in a household. By not allowing dogs to breed freely we have influenced how dogs behave by generally not selecting those with unfavourable characteristics to breed from and tending to select for juvenile and/or highly specific characteristics. By using procedures such as neutering we have influenced the hormonal profile of our companion dogs which has many profound effects on behaviour, such as reducing sexual behaviour. Management of dogs as part of our households also removes the requirement for dogs to perform natural behaviours (though not necessarily the motivation) and training serves to shape the behaviour of animals through reinforcement. Evidence also suggests that when living with humans dogs often override olfactory cues with information gained visually, which would indicate a sensory shift on their part to communicate more effectively with their primarily visually orientated owners.

Given all these influences on the domestic dog, how important is their sense of smell and scent communication to them? How, when and why does it impinge on our generally positive relationships with our companion dogs?

Speaking Different Languages – When Scent Communication Between Dogs and Humans Breaks Down

We have discussed how it is part of a dog's natural behaviour to mark the territory in which its pack resides. However, in a domestic situation there is a fundamental difference – the humans who form part of the 'pack' do not urinate to mark the territory borders or to show their social status. Furthermore they do not appreciate the canine members of the pack performing these tasks at important sites (from the dog's perspective) in the territory, restricting them to specific sites such as the garden. How is this interpreted by the dog? This may depend on the initial human-driven toilet-training process and the relative social status of the dog in the household pack. A puppy, who doesn't yet know better or is unable to help him or herself, can perhaps be forgiven for weeing in the house. But when an adult dog, which the owner believes should know better, releases urine in the house, conflict occurs between animal and owner.

INAPPROPRIATE ELIMINATION

When owners ask for advice on inappropriate elimination they are normally referring to a breakdown in toilet-training. Their dog has begun to urinate and/or defecate in areas the owner finds unacceptable. This doesn't mean that it is necessarily showing abnormal behaviour, just that the dog is going to the toilet in the wrong place. There are a plethora of differential diagnoses for inappropriate elimination. These include certain medical conditions, age-related problems, environmental change, incomplete house training and faulty learning, breed type, inadequate management, substrate preference, submissive urination, over-excitement, fear-related urination, attention-seeking behaviour, status-related behaviour and anxiety problems. The diagnoses related to scent-marking are status-related marking, attention-seeking behaviour and some anxiety problems.

Since inappropriate elimination can be a complicated area to diagnose successfully, it is important that the owner seek professional help in dealing with this problem. The dog should first be examined by a veterinary surgeon to rule out any medical problems. There are many common medical disorders related to urinary problems such as urinary tract disease, congenital or anatomical malformation, endocrine disorders and neurological abnormalities. Problems relating to inappropriate defecation include bacterial and viral conditions, obstructions, food allergies, age-related problems, activity levels and changes in gastro-intestinal motility. If there are no obvious medical reasons the owners can then be referred to a qualified behaviourist.

SCENT-MARKING

Both adult males and females may display this behaviour. To differentiate between scent-marking and other diagnoses it is important to look at the amount, frequency, directionality and locality of the urine deposited. When urine and associated scent secretions are used for communication, only small amounts are usually deposited frequently and directed at prominent places, though low-status animals may pool in environmentally and socially stressful situations. Often the dog will mark after investigating and sniffing several spots. Commonly reported places are the sides of furniture and walls, standard lamps, upright vacuum cleaners, shopping bags and visitors' trouser legs. It is also common for dogs to jump up on to a bed and urinate or defecate on the duvet. There are different motivating factors reported for inappropriate scent-marking behaviour. Confident dogs will often mark as a sign of their status and for territory maintenance. Others may have learned that if they perform these behaviours they get attention, and some dogs may mark as a consequence of feeling anxious. It is, therefore, important that the underlying factors are clearly understood before employing a behavioural modification program.

STATUS-RELATED MARKING AND ATTENTION-SEEKING

Both urination and defecation may be used to scent-mark, which is normal canine behaviour, but inappropriate to owners when it occurs within the home. Among females scent-marking is affected by their hormonal state and they will pass small amounts of urine more frequently when in pro-oestrus, oestrus and post-parturition. Marking behaviour can be reduced by neutering and 'problem' marking can be prevented or at least decreased in over 50 per cent of male dogs by castration. However scent-marking is not controlled exclusively by current levels of reproductive hormones circulating in an animal's bloodstream; hormones produced in other parts of the body play a role and social responses and emotional state may also influence physiological processes that lead to the exhibition of such behaviour patterns. In multi-dog households and free-ranging groups over-marking is common, and more confident females will urinate over urine marks made by males or other females to reaffirm their social status. The dogs which tend to show this confident marking behaviour also tend to show other status-related challenging postures towards other dogs, such as direct staring, assertive body language and vocalising. If scent-marking behaviour occurs within a multi-dog household, it is important to address this problem by reassessing the status between the dogs and treating them accordingly. Greet and treat the more confident dog first and clean thoroughly any urine marks that the others have made. However, status-related marking may also occur in a home where only one dog resides. Commonly a dog will mark on walls or furniture and even a trouser leg in front of its owner. The leg-raising posture on its own may also act as a visual threat display. The dog may also be showing assertive behaviours such as aggressive behaviour if asked to do something, or persistent attention-seeking behaviour such as head-pushing or toy-bringing. If this is the case the owner should re-assess his or her relationship with the dog and gain advice on how to decrease the dog's perception of status, using non-confrontational methods.

Since the smell of urine in a particular place may encourage the dog to over-mark, it is advisable to clean the areas used thoroughly. Biological washing powder is often used and the area wiped with surgical spirit. There are also several commercial furniture deodoriser sprays which clients have recommended.

CASE STUDY

A relevant case study involved a Jack Russell dog called Bob who lived with a recently married couple who, on demand, gave him every thing he asked. They constantly played throw-the-squeaky-toy for him, gave into his food-begging, allowed him to climb all over them persistently and at night he would artfully push the husband out of the bed. As soon as either of them stopped interacting with him he would run into the hallway and defecate on the mat. This made both owners come running and although Bob was told off he still gained a reaction from them albeit a negative one. He would also walk boldly over towards any visitors, lift his leg and urinate all over their trousers. It was this particular behaviour that the owners initially asked advice about.

Several underlying causes were recognised in this case. First, a consequence of this was that he soon learned that by defecating on the carpet he could make his owners react. Secondly, Bob was also urine-marking the legs of visitors when they arrived at the house. The owners were tactfully asked to ignore attention-seeking behaviour and, if necessary, gently push Bob away without looking or talking to him when he tried to initiate any form of contact. They were then asked to solicit all interactions with Bob by first asking him to come to them, to sit down and to wait. He was not allowed to have a fuss, a treat, his dinner, lead on, or a toy, until told. Bob soon learned that to get nice things he had to sit down calmly and wait first. When visitors arrived they were asked to ignore him and for a few weeks he was kept on a lead in this context. All marks were cleaned using biological washing powders and a deodorising spray. He has not marked since.

ANXIETY PROBLEMS

Some dogs, typically of low status or a nervous disposition, may mark as a consequence of feeling anxious or unsure of a new environment. There are many reported cases of dogs which urinate or defecate when they first enter new surroundings. Once they have performed the deed they rarely repeat it. To discourage this behaviour a deposit of their smell can be laid prior to the dog entering, by wiping an old towel/blanket that the dog has been sleeping on around the walls and furniture. It would also be advisable to have the dog with you and supervised until it has gained in confidence.

Some dogs find it hard to cope when left, and may show anxiety problems such as increased vocalisation, destructive behaviour and elimination problems. Usually these involve full bladder elimination of urine and no scent-marking as such. However, it has been suggested that some dogs scent-mark to increase their feeling of security. Therefore, the separation-anxiety problem needs to be addressed before the marking will decrease. Not all problems are mutually exclusive and can figure a multitude of underlying causes and it is important that each dog should have a medical check-up before being referred to a qualified behaviourist. In some cases appropriate drug therapy may be recommended by the veterinary surgeon. In these cases and concurrent to the behavioural modification program the dog appeasement pheromone (DAP) may be of some use.

ROLLING IN SCATS

Rolling is a normal canine scavenging and hunting and possibly status-related behaviour, which is generally ignored unless the odour is unpleasant to humans – such as rotting fish. It is fair to say that a dog's motivation to pursue such odours is usually very high so if the owners do not wish their dog to smell like a fish they must be closely observe their dog's behaviour to try and anticipate rolling behaviour and work on the dog's recall. After a bath, rolling is likely to aid manipulation of the dog's scent glands and spread their scent over their body once more. To stop them rolling in mud directly after a bath it is advisable to keep them indoors and give an alternative low-odour substrate to roll in, such as their bedding.

ROAMING

Intact male dogs roam more frequently than castrated ones and will sniff out sexually receptive females over long distances. The obvious dangers include being injured or killed on the roads and causing third party injuries or fatalities. Unsupervised dogs also pose a health hazard and fighting can also break out with other roaming males. So it is important that this problem is addressed. Since roaming behaviour is usually hormonally modulated castration will help to decrease this behaviour by about 90 per cent. However, this behaviour also has a learned component, and the longer the dog has been showing this behaviour the less likely castration will help to reduce it.

If, after neutering, the dog is still an ingenious escapologist, the owner should make sure that it is sufficiently exercised and its attention needs met and that it wants to stay with the owners rather than being made to. Eight-foot-high fencing sunk into concrete and surrounded by razor wire may physically stop the dog from running away, but it will not stop the dog from wanting to. A qualified behaviourist will be able to help with this problem.

STEALING SCENT-IMPREGNATED ITEMS

Many dogs pinch scent-impregnated items, such as tissues. This can include the owner's underwear which the dog may refuse to give back. Why a dog would choose these items over and above others we can only surmise. However, these items of clothing do carry a relatively large amount of the owner's scent. They are also a handy size to fit comfortably into a dog's mouth. When taken, these items procure an explosive reaction from the embarrassed owner. Therefore, the dog quickly learns to make the association between the item and the reaction of their owner, which may be interpreted as play, such as tug-of-war. The initial motivation to take these items thus changes from a natural interest in pack odours to a learned association, which gains the reward of attention. One solution to overcome such an attention-seeking motivation is to dissociate the item with the reward. Instead of becoming all flustered and chasing the dog around the house, simply ignore the dog and fetch something 'better' (determined by your excitement and interest) such as the dog's favourite toy or treat. Become very interested in this 'first prize' and the dog will quickly lose interest in its now 'second prize' and will come trotting over. Do not try to force the item away from the dog. Simply ignore and wait until the item is dropped. Carefully pick up the soggy underwear and sometimes ask the dog to sit, wait and take what you have. The dog will learn that to receive nice things it has to sit and wait first and that pinching items of underwear does not achieve the desired result of attention. Some dogs will take items impregnated with the owner's scent to lie on, or lie on those that are already available at times of stress, such as during separation anxiety. This acts as a maintenance stimuli that increases their capacity to cope.

Conclusion

Scents are important to dogs, much more so than to us. They convey messages relevant to their (ancestral) survival and social behaviour. Such messages are not obvious to us humans, as we do not have the same ability or motivation to respond to them. Unfortunately this can and does sometimes lead to a breakdown in dog-human communication. Scientific studies of the behaviour and physiology of dogs and the experience of behavioural counsellors in deciphering how to deal with dog-human misunderstandings has helped to unravel some of the secrets of the canine world of scents. However, there is still much to learn which requires us to keep our eyes, ears and most pertinently our noses open to what our dogs are telling us.

7
The Relationship between Emotions and Canine Behaviour Problems

David Appleby and Jolanda Pluijmakers

To explain the development of behaviour problems and know how they can be prevented and treated, we must first understand how they are influenced by emotions. Emotions are experienced either positively or negatively. They result from a match or mismatch between events in the environment and the animal's interests. A mismatch gives rise to a negative emotional state, such as fear or anxiety, which causes the individual to act in a certain way in order to cope with a threat to its physical or emotional homeostasis. The emotions experienced are a physical reaction, which takes the form of an increase in the parasympathetic (relaxation) or sympathetic (arousal) part of the autonomic nervous system and corresponding decline in the other, as discussed in Chapter 2. If the increase is in the sympathetic, the state of arousal provides a fast mobilisation of energy. Hormone release strengthens the autonomic response by an extra increase of the blood flow and making more glucose available to the muscles. The behavioural response consists of the muscular movements, postural changes, facial expressions and sounds appropriate to the situation that elicits the emotion.

The function of emotions is to increase an individual's chance of survival. This is most clearly illustrated by the autonomic and hormonal reactions that prepare the body for actions, such as flight or fight when faced with a threatening stimulus. For our and the dog's pre-mammalian ancestors this was all emotions consisted of; however, other responses and functions evolved. Expression of emotional states occurs through behaviour, body language and sounds, all of which serve a useful social function, with survival as the end-gain. Such expression influences the other individuals' behaviour since it will then be easier for them to judge what the individual expressing the emotion is likely to do. For example, a threatening growl and body language, indicative of arousal and directed at an individual that tries to take a valuable resource, may cause it to withdraw. By communicating emotional states the stability of social groups is enhanced, the potential for conflict decreased, and the chance of survival increased. An example of this might be when distress vocalisation is triggered because a stimulus which the individual is dependent upon is not available, and this results in reunion with the other group members.

Emotions are subjective, which means that although the process has the same characteristics for all members of a species, it does not result in the same outcome for each individual. How an animal is affected by a situation is influenced by many factors, such as genetic and environmental influences on the development of the brain, past learning experiences and the complex information processing procedure within the brain. This last is influenced by how different parts of the brain respond to different stimuli and contexts. From this it follows that the coping response, through which an individual minimises an aversive effect of a stimulus or context, may differ enormously.

Genetics

Recent research of the heritability of personality in identical twins shows that personalities are established by both genetics and environmental factors. The sensitive period of behavioural organisation, discussed in Chapter 2, is an example of how genetically predetermined biological development combined with environmental factors, such as learning experience, have a profound influence on the emotional development of an individual.

That an emotionality can be inherited has been shown experimentally in the parallel breeding of two lines of Pointers. Dogs from Line A were confident in the presence of people and remained so. Dogs from Line E were shy of people. Puppies that were bred from dogs in this line showed severe aversion of human company. Cross-bred descendants (one parent from Line E and one from A) were initially confident in the presence of people but subsequent generations of this cross resembled line E and were shy of people. This experiment suggests that in these lines the inheritability of nervousness was dominant. This is important because the genetic potential for nervousness and other emotional states helps to determine how an animal will react in a situation.

By observing the emotional expression of his children and comparing them with people living in various isolated cultures around the world Darwin (Darwin, 1872/1965) was able to conclude that there is an innate species-typical repertoire of emotional expression.

Although all members of a species inherit the full repertoire of behaviour from their ancestors, in dogs, as discussed in the chapter about their ethology, the capacity to perform particular tasks has been increased in some and allowed to diminish in others through selective breeding. Therefore some behaviour is easier to trigger in some breeds than in others.

The variability of behaviour from breed to breed was shown in research conducted in the 1960s when a number of breeds were reared in identical circumstances and tested at different ages for their behavioural performance. Fox Terriers, for example, were described as aggressive and non-fearful, Basenjis as aggressive and fearful and Beagles as non-aggressive and non-fearful. If breeds are genetically predisposed to display patterns of behaviour, it follows that the type of problem behaviour displayed by dogs will often be a result of their genetic predisposition. For example, a Border Collie has such a strong drive to herd that it will not only herd sheep but any livestock whose movement stimulates the response. When given inappropriate learning opportunities, it may also be stimulated to chase cars, cyclists and joggers.

The genetic predisposition for the performance of behaviours that are suggested to have no obvious function and can be triggered by stress provides a further illustration. For example, a Dobermann may suck its flanks whereas a German Shepherd Dog may chase its tail.

The chemical processes associated with the emotional state of the bitch may affect the whole litter. For example, a bitch which is stressed during pregnancy is more likely to give birth to emotionally reactive puppies. Within a litter individual differences are further refined. There are differences in some types of behaviour and behaviour problems displayed by males and females. It is suggested that already within the womb gender differences in behavioural potential may occur due to the position the puppies lie in. For example, a bitch that lay between two male puppies during pregnancy may have some masculine characteristics when compared to a bitch which lay between two bitches.

Learning

A few stimuli, typically those that present a significant, life-challenging situation, automatically produce an emotional reaction. Other stimuli do not cause a spontaneous emotional response, instead it has to be learned through classical conditioning during which the emotion becomes associated with a neutral stimulus. If the stimulus is associated with an evasive experience the individual will become frightened when it encounters the conditioned stimulus or a comparable situation again. As every individual strives to maintain a state of emotional homeostasis, they will display behaviour to try to avoid or remove this negative emotion. When they have learned an effective coping strategy that minimises the aversive effect or avoids the stimulus, the emotional response will no longer occur.

OPPONENT PROCESSES

Emotional states are automatically opposed by mechanisms of the central nervous system, which reduce the intensity of both pleasant and unpleasant feelings. Following the introduction of something pleasurable, a positive emotional response begins and quickly rises to a peak (State A). It then slowly declines to a level where it remains provided the quality and intensity of the stimulus are maintained. When the stimulus is removed, the reaction gives way to a qualitatively different type of reaction (State B), which reaches its own peak of intensity and then slowly disappears with time and goes back to normal. These reactions are opponent, so if A is pleasant, B is aversive; conversely if A is aversive B will be pleasant. The B-process will acquire more strength if frequently elicited. It will also rise to a higher peak and take longer to decay than the A process. In addition the effect of the A process decreases through repeated stimulation.

The effects of frequent State A stimulation may be at work in dogs displaying attention-seeking behaviours such as stealing. By giving a lot of attention to a dog, pleasure (State A) is created. Repeated stimulation of State A results in an increasing craving for attention when it is not available (State B), and more attention is required to return the dog to State A and so on. The craving can become so strong that the dog may learn the strategy of stealing objects, such as the remote control for the TV, purse, the owner's glasses, etc. when the owner is preoccupied, because it finds that the owner will always respond when it picks them up.

How Emotions Are Processed

Emotions are commonly experienced when a stimulus, context or behaviour activates memories. These memories differ from factual memories because they contain response information and activate the associated systems resulting in an experience of the emotion. What takes place between the perception of a stimulus and the response to it can be described as data-processing within the brain. This process is assumed to be present because the response to a stimulus varies.

The process involves a number of stages, which are summarised below:

1. Assessment of the stimulus, for example, is it familiar or unfamiliar? How do the properties of stimulus compare with other stimuli?
2. Assessment of context: The context is evaluated to assess the potential to remain in control of emotional homeostasis in the presence of the stimulus/event.
3. Evaluation: Based on information from previous stages, the importance of the stimulus/event is calculated, which determines the priority it is given. If it is high, ongoing behaviour will be disrupted.
4. Behaviour plan: An action plan is generated based on information from previous stages.
5. Physiological changes and selection of behaviour.
6. The behaviour is performed.

In contrast to what this might suggest, the process is not an isolated, linear one. The recording of information and the evaluation of its relevance are continuous. As long as an organism is awake it will scan/explore its environment and compare the stimuli it encounters with its interests, which are not static.

It is normal for an emotion to result from the process described above. However, parts of the process can be 'left out' or overtaken by other processes, and they can be interrupted at any time. For example, context evaluation may be skipped in response to sudden, intense or unexpected events in which there is no time to evaluate. Everything between context evaluation and overt behaviour may be left out if the individual has learned a response that is strongly established or is an innate response.

Anxiety, Fear and Phobia

Anxiety, fear and phobia result from a loss of emotional homeostasis, which in turn causes activity in the sympathetic autonomic nervous system. Variation in behavioural signs is dependent upon the characteristics of the individual, the situation and the intensity of the emotion experienced. Some dogs become restless, pace, tremble, pant or salivate when anxious. Others may bark, whine or seek comfort from maintenance stimuli, e.g., by trying to hide where it feels secure or seeking proximity to the owner. When fearful they might freeze or try to escape by flight e.g., digging into locations, scratching at doors or windows. If necessary a dog may display defensive behaviour in an attempt to escape from the situation and return to emotional homeostasis.

Not being able to predict the outcome of a situation is frightening. Inability to predict makes it difficult or impossible for an animal to organise its behaviour so that it can respond appropriately. As a consequence of a sense of loss of control, its emotional homeostasis may be disrupted, and it will become anxious or frustrated. This can lead to the development of behaviour problems ranging from major to what might be considered nuisance behaviours, such as whining or jumping up. However, apparently mundane problems caused by an animal not being able to predict its environment can develop into something more serious.

A dog's need to be able to predict its environment and have ways of controlling how it is affected emotionally also applies to the social group; in most cases this will be the family and any other pets it lives with. If owners are inconsistent in the way they interact with their dogs, at worst appearing to swing from a 'Jekyll' to a 'Hyde' personality because they don't understand their pet's perception of their behaviour, the inability to predict its owners and know how to respond may result in it becoming insecure, a potential consequence of which will be defensiveness. This is especially true if it is subject to higher levels of stress through reduced ability to cope because poor behavioural organisation and/or members of the family are inclined to be punitive or heavy handed.

Here is an example of what has been discussed above. If a dog is never fed tit-bits from the owner's plate it will not expect anything and will not look for it. However, if some others in the family sometimes feed the dog from their plate they create an expectation that it will happen again and so the dog learns to sit and watch people when they are eating – just in case. If it takes longer than it expects for someone to feed it, he might try a reminder that he expects to get something by pushing under the person's arm. If that person does not react immediately, the frustration it experiences will make it try harder and push more or try another strategy, such as barking. Not such a problem you might think, but the dog's ability to predict its environment and know how to respond will take a turn for the worse, if it then finds that one or more family members, perhaps not those who did the original feeding, tell it off for 'begging'. The extent to which it is affected emotionally will depend upon the degree of threat the dog perceives itself to be under and its confidence in its capacity to cope. It may try to cope by avoiding its owners in this context, giving them what they intended, or by becoming defensive. Any telling-off for being defensive that follows will only confirm to the dog that it was right to fear its owners. In response to the owners' unpredictability in this context the dog may become defensive in others where it perceives itself to be confronted, just in case the owners turn nasty.

We have already seen that an animal can inherit a predisposition to fearfulness. We have also seen that early experiences during the sensitive period of behavioural organisation and the formation and development of their maintenance set have a profound effect on the animal's capacity to maintain emotional homeostasis in a changing environment and when it encounters novel stimuli.

Learning experiences through classical and instrumental conditioning are relevant to the development of anxiety, fear and phobia, its generalisation to other situations and any increase or decrease in the intensity of the animal's emotional response. For example, a dog which is always friendly to other dogs may, as a result of being bitten by a large black dog, start to avoid large black dogs. This behaviour can easily generalise to all types of dogs through emotional responses being triggered by dogs which look almost the same size and colour, than ones that are less similar and so on. If the dog has several more unpleasant experiences, it may become phobic of other dogs and display an intense and immediate emotional reaction every time it sees another dog. Conversely, if the subsequent experiences with other dogs are positive, the same dog may learn to overcome its fear.

Some stimuli have a higher genetic potential for triggering an immediate and intense fear response. The large number of dogs which suffer from noise phobia suggests that an example of such a stimulus for them could be the fear of sudden loud noises – fireworks or gunshots. When this stimulus is encountered there is often an immediate and intense emotional reaction before the brain has identified what the noise is, where it is coming from, if it indeed is threatening to the individual and what the best behavioural strategy would be in this situation.

All these steps in the emotional information processing procedure are initially skipped because in a life-threatening situation there is no time for extensive and relatively slow evaluation. A fast and immediate reaction may substantially increase the chances for survival. Stimuli present at the time of the bang are easily associated with the negative experience and can subsequently trigger fear because of that association. If, in addition to this, the animal is able to anticipate that something frightening may happen because of previous experience, it may become anxious, especially if it does not have an effective means of coping.

Here is an example of how this can influence an animal's behaviour. Some dogs become reluctant to go out to relieve themselves before their owners go to bed, or go for a walk, after dark because they are concerned that fireworks may re-occur. Except for certain times a year, we can predict that it is unlikely that fireworks will occur, but for the dog darkness allows it to anticipate that they might, and this association will only extinguish over a period of time during which darkness is not followed by the onset of fireworks. However, the next time they occur at night, the dog's capacity to predict its environment and maintain control of its emotional state is further reduced, and further still if they occur during daylight. Remaining indoors is no guarantee that the dog will be able satisfactorily to control its exposure to the bangs and emotional state, because to a lesser extent they continue to occur. For some this reduction is sufficient to enable them to cope. Others have to use strategies to control their sense of security, such as trying to get close to their owners or hiding in den-like locations. When the owners are not at home, fear of the noises or the expectation that they may come can result in separation problems, such as digging into places to hide or escape by any means. If none of these works, the dog experiences complete loss of control: it may lose toilet control and exhibit other symptoms of anxiety and fear.

The analyses of the context through information processing in the brain can lead to emotions of anxiety or fear. Fear or anxiety results from a loss of control that is experienced when an animal loses too many stimuli from its maintenance set or the changes in its environment make it more reliant on its maintenance set. For example, a dog that has grown up with an older dog, and has learned to depend on it as a means of maintaining a sense of security on walks, may behave differently if it has to be walked on its own. It may stay much closer to the handler, check more often to see if s/he is still there, scan the environment more intensely and startle more easily. If, in addition to being out 'alone' without its giant maintenance stimulus (the other dog) it is taken to a strange environment where the dog's capacity to predict and control is further impaired, it may become stressed and start to pant, pace or tremble. Minor stimuli it would normally not react to may cause an intense emotional response, which may lead to it becoming extremely reactive, and barking, growling or trying to hide behind the owner's legs. It is for this reason that when introducing new stimuli or environments the composition of the maintenance set should be evaluated carefully in order to prevent the dog from developing negative associations with the stimulus or the environment, and at the same time preventing it from being too dependent on one stimulus.

Separation Problems

Problems that occur when a dog is left at home alone constitute a significant portion of the behavioural specialist's caseload. Until the 1990s the term 'separation anxiety' was commonly used for all separation-related problems. However, there are many causes for separation problems that are not related to anxiety. One type of separation problem, namely 'separation anxiety', has been defined more specifically and is often described as problematic behaviour motivated by anxiety that occurs exclusively in the owners' absence or virtual absence. Rather than being exclusively caused by being left unattended by the owner, as suggested by some authors, separation anxiety can be defined as apprehension due to removal of significant social or non-social maintenance stimuli which can, for example, be people, other animal(s) or familiar surroundings.

Fear and anxiety and the opponent process theory play a considerable role in separation anxiety. Commonly reported symptoms, such as destruction, vocalisation and house soiling, are indicative of a loss of emotional homeostasis, and some symptoms seem to be associated with attempts to regain access to the lost maintenance stimulus (see table). The opponent process affects the dog's emotional homeostasis because comfort due to social contact (A Process) is replaced by negative emotions due to loss of sense of security and control when the dog is left alone (B Process). Repetition of these contrasting emotional states results in attention-seeking and neediness when the owners are at home and as a consequence a really emotional state occurs when the dog is left alone.

Based on this last definition of separation anxiety given above, the composition of maintenance sets can be divided into three groups, A, B & C.

Dogs in Group A are those that have remained dependent upon the mother figure or a substitute for them, such as the first owner. This typically occurs where, as a puppy, the dog was not able to explore and develop some independence. This may be caused if the owner is inclined to encourage close contact rather than independent behaviour, where the puppy has been unwell, etc.

Dogs in Group B are those that have developed a broader maintenance set of social and non-social things through learning, but their emotional homeostasis can still be upset if one or more things they heavily depend upon are removed from the maintenance set, and/or their confidence to cope is reduced by a perceived or actual threat to its well-being.

Dogs in Group C have such a broad maintenance set that they can cope with the removal of stimuli from it, because enough remain to enable it to stay in a positive emotional state. However they can still be made fearful by the presence of aversive stimuli or other experiences. The more complete its maintenance set, the better able a dog is to cope.

Dogs can move between groups B and C. For example, a dog may have a broad maintenance set (group C) and not be dependent upon an owner, but if an owner stays at home for a long period of time during redundancy or maternity leave, his or her availability, the quality of the relationship and the resultant emotional association may result in the owner becoming overly important in the maintenance set (group B). A dog belonging to group C may move to group B after moving house with his owners. Having lost a lot of its non-social maintenance stimuli and being surrounded by unfamiliar stimuli can create a temporally inadequate maintenance set leading to separation anxiety. With time, however, the dog will probably rebuild its maintenance set, making it possible to cope with the removal of maintenance stimuli again, and move back to group C.

The symptoms seen in dogs from these three groups are likely to vary, as shown in the table on p.141.

There are other reasons why separation problems occur when dogs are left unattended, the symptoms of which are the consequence of an emotional state. Some problems occur because the dog is not able to anticipate that they are expected to stay, resulting in feelings of frustration and annoyance when they can't leave with the owner. This may result in problems such as barking in a manner consistent with these emotions or giving the cushions a good shake. It is easier for a dog to know that it is going to stay behind and thus avoid these problems if its owners wear different clothes for walking it and going out without it. Even a pair of Wellingtons or a hat only worn when it is going to be taken out will help. The dog's lead being left undisturbed as the owner gets ready to leave and being left behind once they have gone should cause the dog to expect to stay at home. However these clues are missing if the owner hangs the lead with the family's coats so that it rattles every time someone gets a coat, or keeps the lead where the dog cannot see it. It is preferable for the lead to hang on its own hook in an area of the house where the dog can always see it and can learn that that if it remains undisturbed it is going to stay behind.

Group	*Time of Onset*	*When owner present*	*Behaviour during departing and greeting*	*When owner absent*
A	From puppyhood Symptoms start within minutes of being left, often immediately after the departure or absence of the social stimulus e.g., A person, the dog is specifically dependent upon.	Organisation of all activity around one social stimulus. Follows about the home. Needs physical contact, e.g., Wants to be next to a specific owner. High need for attention/affection. Tends to stay close to the social stimulus rather than explore on walks.	Distressed when separation is anticipated, e.g., Trembling, shaking, depressed. May attempt to prevent departure. Over excitement when owner returns.	Destruction orientated towards doors and windows that give access to the direction by which the social stimulus left. Destruction of items impregnated with the owner's scent, e.g., Bedding, remote controls, papers, shoes. Vocalisation consistent with distress/relocation.
B	Starts after removal of one important stimulus, several less important stimuli from the maintenance set, and/or because the dog has an increased need for the maintenance set e.g., due to a new fear such as rehoming, moving house, left in other room. Only when the dog is left in circumstances where its maintenance set is inadequate.	Dependence on the presence of one or more social stimuli and/or non-social stimuli, e.g., location in the home. Dependence on social stimuli can increase if separa tion unpredictable. May show dependent behaviour if there is an increased need, e.g., if bangs occur.	Distress during departure and excessive greeting. May attempt to prevent departure. Departure may result in agitation or depression.	If dependent on social stimuli destruction occurs during attempts to gain access to them. If the dog is sufficiently dependent upon inanimate stimuli, it may dig so as to hide or get into rooms shut off from them, such as the owner's bedroom. If fearful they may escape, by any door or window. Vocal due to distress/relocation but may not occur if caused by fear or if the dog is not over-dependent upon absent social stimuli. Defaecation and urination may suggest a sense of loss of control.
C	Problems start after an experience that causes fear or phobia that may be associated with the absence of the owners and triggered by it.	No dependency upon specific stimuli. May be fearful of the stimulus when owner is present. The extent to which this occurs may be less because the maintenance set is more complete, behaviour is more ordered and the dog has a greater sense of control.	Distress signs can develop if owner absence is associated with stimuli that cause fear.	Defecation and urination can occur due to emotional state associated with loss of control. Dog may try to cope in a disarranged way or an organised way by trying to escape or hide.

Aggression

Aggression is almost certainly the most acute and threatening behaviour problem for the dog-human relationship. Aggression may be directed towards other animals belonging to the same or another species, familiar and unfamiliar people or objects, such as cars, and serves a variety of functions in an animal's life. In its natural setting aggression has value for the survival of the individual and species. However in the domestic dog the behaviour is often misinterpreted because of judgement based on its observation that does not take into account how the behaviour developed and is influenced by factors and processes, such as genetics, emotional development and learning experiences. For example, a dog overtly barking and lunging at another dog may be labelled as 'dominant' by the casual and ill-informed observer. However, the cause of the behaviour could be fear. Although dogs may initially try to avoid or escape from what they are frightened of, they can develop a coping strategy of defensive behaviour to keep it at a distance. The contrast between the emotions caused by what they perceive as aversive and the sense of relief when it goes away results in the reinforcement of the defensive behaviour, and it becomes increasingly overt. As a consequence of the apparent success, most stimuli go away of their own volition, the dog's confidence is increased and this reduces its fear in that context. This can result in body language that does not suggest fear, leading to a wrong assumption about its motivation. A more accurate approach is to describe what we can observe of the animal's behaviour. For example, instead of saying a dog displayed a particular type of aggression, it would be more accurate to describe what it actually did, e.g., it growled, barked, bit, approached or withdrew, etc. We can separately consider its emotional state from one situation to another through analysis of the available history.

Most of the available literature tends to categorise aggression according to perceived motivation. Examples include fear, dominance, redirected, possessive, protective and predatory. Looking at aggression from a biological point of view, this extensive and confusing list can be reduced to resource-related aggression, fear aggression and aggression resulting from a physical cause.

Social animals spend a vast amount of time together, and conflicts may be inevitable because resources are commonly limited and not every animal can exploit them equally. They must communicate with each other to gain access to and use of them. To reduce the possibility of fighting, species-specific genetically programmed ritualised signals are displayed and increase an animal's and the groups' chance of survival. Vocalisations, facial expressions and body postures communicate the emotions of animals involved in the conflict. Various factors influence the potential for aggressive behaviour. These include temperament, learning from previous encounters, observational learning, e.g., from the mother, hormones, age, sex, size, distance and degree of social inhibition.

Aggressive behaviour, caused by competition for resources and directed towards pack members in the natural environment or familiar people or dogs living in the same household, has until recently been called dominance-related aggression. Descriptions of this behaviour include growling or biting when competing for or protecting resources such as a bone, a favourite place on the sofa or food bowl.

In simple terms a sense of being able to control resources can be expressed as an equation of *Resource Holding Potential* (RHP) + *Value* over *Cost*. This comes down to: Do you normally get your own way? + Something you want – Any concerns about the ability of an individual wanting the same thing to stop you and cause you harm. If the dog thinks it is stronger than the other individual, really wants the resource, and does not think the other individual will harm it, it is likely not to feel inhibited by them, and will control the resource. If two individuals keep having incidents where this equation comes into effect, they should reach the point where one can anticipate who will win and in what contexts this is likely to occur. This will result in a ranking system that avoids challenge and conflict.

Ranking systems have evolved in pack society because fighting decreases when everybody knows where they stand. Problems tend to occur in the human/dog relationship when owners convey signals that unwittingly cause their dog to develop a sense of control over resources and become uninhibited in their relationship with them, increasing the potential for agonistic behaviour, of which competitive aggression is one aspect. For example, an owner who readily hands over the food they are eating when their dog demands it may be developing in it a sense of control over the resource of food in that context. Of course this act on its own is unlikely to upset the dog's perception of the appropriate ranking order but if it perceives itself as being able to control resources in a number of situations it may become uninhibited about challenging its owners in other situations. The importance of control is shown in some research in which it was found that dogs are more likely to develop lack of inhibition in homes where their owners acquiesce to their demands (O'Farrell, 1995).

Contrary to popular belief fear and anxiety may play an important role in status-related aggression, for example, fear of loss of control of those resources that have become important for the dog's emotional homeostasis. Behaviour modification programmes introduced to cure a hierarchy problem, which often involve the owners denying access to resources the dog cares about, can result in depression and withdrawn behaviour. It is thought that this is a reaction to failing to get anticipated pleasures such as portions of the owner's food or to lie on the sofa. However, once the dog has adjusted to the new situation and its expectations are lowered, this tendency stops. Nonetheless, it is necessary to reduce expectations so that aggression does not occur as a consequence of frustration when the dog's expectations of being able to have what it wants when it wants it are not met. To prevent a negative emotional state of depression occurring, owners can request their dog to earn everything it wants or is given, by responding to signals to sit, lie down, etc. before the resource is given – thus effectively maintaining the owner's control of the resource.

It has also been argued that many instances of aggression towards people in the dog's social group, perhaps the majority, are not a status problem but related to avoiding an aversive outcome during an associated negative emotion. During social conflict dogs that have become uninhibited may learn to use defensive coping strategies to avoid an aversive event more readily than inhibited (submissive) dogs, which are more likely to choose to flight or freeze as a way of coping.

Anecdotal evidence suggests that dogs presented for the treatment of status-related aggression are often found to be nervous and constantly seeking attention from their owners, and are described as being generally fearful. When making detailed analyses of the bite incidents, the display of aggression can often be traced to a stressful situation, such as the sound of a loud noise occurring just before the incident. Anxiety may be caused by other chronic environmental stress or a lack of predictability in social relationships or physical environment. If owners add to the stress of fearful or anxious dogs through apparent confrontation, even if not intended, their behaviour may be somewhat disorganised, resulting in a range of possible behaviours that includes excitability, failure to respond to their wishes and defensive aggression. These may be misinterpreted as 'dominant' behaviour.

Physical causes for aggression include physical and perceptual ability, pain, endocrine disorders, liver or kidney disorders, brain disease and the influence of diet. The strong link between medical conditions and some behaviour problems explains why the APBC strongly believes that best practice requires that medical causes for behaviour problems are excluded or identified and treated before behaviour therapy is introduced. Failure to do so could be to the detriment of the welfare of the animals in our care and therefore all behaviour problems are seen on veterinary referral.

There are two types of behaviour, redirected and predatory, that are often categorised as aggression, that should probably not be. Redirected behaviour occurs when a dog is in an aroused state, which may or may not involve aggression, but cannot get to what has caused its arousal and it redirects its arousal as aggression, typically on to what has distracted or frustrated it. Predatory behaviour is not an aggressive act but associated with obtaining food, and different areas of the brain become active.

Conclusion

Behaviour is the means by which an animal maintains emotional homeostasis. From the outcomes of its behaviour it learns to display the most effective response for that and similar situations to achieve this goal. Many behaviour problems, including minor ones, stem from disruption of emotional homeostasis and/or behaviour that has become disorganised and ineffective as a coping mechanism, or is perceived by the dog to be effective but is inappropriate for our requirements as pet owners and society in general. Therefore, it is essential that the emotional states of an animal are properly understood and what is learned included in our approach to preventing and treating behaviour problems.

8
Feline Behaviour Problems – the influence of natural behaviour

Sarah Heath

Introduction

In a domestic environment many of the natural feline behaviour systems are compromised and cats find themselves living in groups of unrelated individuals, being made to share important resources and denied the opportunity to hide or retreat from situations of potential conflict. In addition, their human companions place social demands on them that are at odds with their natural behaviour and often fail to provide for many of their basic instincts. Such constraints on normal behaviour result in stress and tension in our feline companions and, in many incidents of reported behaviour problems, an understanding of feline ethology not only helps to explain how and why the more common problems develop but also offers practical methods for dealing with them.

Feline Society

Although the traditional image of the cat as a solitary creature is not entirely accurate, it is important to remember that much of feline behaviour is based on individual survival, and many fundamental behaviours, such as feeding, hunting, resting and eliminating, are performed in a solitary context and have no social significance. Feline society is based on co-operative groups of females who are related to one another and live together in a mutually beneficial environment, which supports the successful rearing of kittens. Males are usually excluded from these social groupings and live their lives as solitary individuals who only venture into the main social context at times of breeding.

Levels of hostility within the related groups are low but intrusion by unrelated individuals poses a potential risk to offspring and is poorly tolerated, so aggression to strangers can be intense. However, physical aggression carries with it the risk of injury and in a society where ultimate survival is an individual responsibility it makes sense to avoid situations which could result in a decreased ability to take care of oneself. For this reason overt aggression is minimised by the use of elaborate distance-maintaining behaviours which are designed to keep strangers at bay and discourage social interaction with individuals outside the social group. These signals include postural and vocal communication, marking behaviours, such as urine spraying, and elaborate use of eye contact and facial communication, including ear positions.

Within this social context the reliable identification of fellow members of the social group is vitally important, and affiliative behaviours, such as mutual grooming and mutual rubbing, are used to cement relationships and to exchange scent signals. Indeed the mixing of scents results in the formation of a group social odour which reassures individuals that the social group is stable and enables them to relax in close proximity to one another.

The Challenge of Multi-cat Households

Over recent years the cat has increased significantly in popularity and at the same time the number of multi-cat households has also increased. Some of these households are made up of sibling pairs, mothers and offspring, and other combinations of related individuals, but in many cases there is no such basis to the feline community and cats are being expected to live with total feline strangers. Owners often acquire the cats as companions for one another and when problems of inter-cat aggression begin to surface within the household they are genuinely dismayed and perplexed by the situation. However, natural feline ethology provides no basis for toleration between unrelated felines and while the high proportion of neutering in the domestic population undoubtedly reduces hostility between strangers, and enhances the chances of feline integration, it by no means guarantees it.

In some cases the aggression between housemates may be manifested as outright physical confrontation and the cats may be taken to the local veterinary practice with torn ears and puncture wounds to prove it, but feline tension can also result in more subtle signs of unease and can even contribute to other behavioural issues such as indoor marking, over grooming and even alterations in appetite.

Dealing with Aggression between Cats in the Household

One of the most important steps in dealing with problems of aggression within a feline household is to ensure that each individual has ready access to all the important resources, such as food, water and resting places. Just because cats live in the same human household does not mean that they belong to the same social group and it is not uncommon for multi-cat households to be divided into smaller subunits of individuals, which do not communicate across those social divides. If this is the case, it will be necessary to ensure that there are enough feeding stations, watering holes and latrines for each of the social groups to exist as an individual unit, and attempts to make these groups share these essential commodities will invariably result in social tension, if not overt aggression.

When physical space within the home is limited, the amount of available territory can be increased by paying attention to vertical space, and the installation of resting platforms and provision of bedding on tops of wardrobes and kitchen cupboards can significantly increase the space available. Such structures will also increase the opportunity for cats to retreat and hide when they perceive themselves to be in danger, and such provision for the expression of natural coping behaviours will assist in reducing stress and increasing feelings of home security. In addition cats which are provided with such escape routes will be less inclined to use self-appeasement behaviours such as over-grooming or excessive eating to deal with their stress, and thus passive symptoms of inter-cat tension will also be reduced.

When cats have previously co-existed in harmony but have started to display hostility toward each other, it is useful to determine the reason for the breakdown in the relationship. For some it may be the result of disruption of their social grouping by the entry of another feline into the home while for others a breakdown in communication may follow a brief period of separation, for example when one of the group is taken to the veterinary practice for treatment.

Regardless of whether the cats have previously co-existed or are being introduced for the first time, integration is unlikely to be successful from a position of hostility and a period of total physical separation can be useful in order to allow scent integration through the mixing of feeding utensils, bedding and toys. In this way the cats can come into contact with each other without any direct form of threat and once they can accept the scent of their housemate without any signs of hostility the owner can then begin the controlled process of introducing the cats in a face-to-face situation. Patience will be the main requirement at this stage and any temptation to rush the introduction will need to be resisted. During the introduction process the overriding aim is to improve the cats' perception of security within the home and the use of pheromone therapy has been shown to be very beneficial in this context. Cats use facial scent deposits to mark their territory and to surround themselves with reassuring signals and synthetic analogues of these signals (marketed as the product Feliway®) are now available which can assist in relaxing feline inhabitants and signalling to them that their home is safe and secure.

Pressures from Neighbourhood Cats

Multi-cat households are not the only implication of increasing feline popularity. Social tension within local neighbourhoods is also heightened by the increasing population density in urban areas. In feral situations the size of feline territories and the density of feline inhabitants will be dictated by the availability of vital resources such as food and shelter, but in domestic situations each cat has both of these resources provided for it by its owner and therefore relatively small geographical areas can sustain relatively high numbers of cats without obvious problems. However, such populations are often very unstable and any minor challenge to their stability, such as the introduction of a newcomer, can result in significant levels of inter-cat aggression and an increase in fear-related behaviour problems in many of the individuals.

In the majority of cases of aggression between cats in the neighbourhood, it is the owner of the victim who first seeks professional advice and, without co-operation from both sets of owners, the prognosis for reform is often guarded. Where aggression primarily occurs outside in the gardens and beyond, management will be aimed at decreasing unwanted encounters between the cats. Effective time-share systems, where owners allow their cats access to outdoors at different times, can work well.

Increasing the feline appeal of each cat's home is also a useful technique and ensuring that all the cat's behavioural needs are catered for within the house and garden can decrease their desire to venture further afield and so decrease the incidence of interaction with neighbouring individuals. However, in some cases, the victims are not venturing far from home and the aggressors are actively pursuing them, sometimes even into their own houses. These situations are far more difficult to deal with and when cats proactively attempt to increase the size of their territory it is important to ensure that vital resources are not being compromised in some way in their home, for example, by competition from housemates or by a basic lack of provision for their behavioural needs. Once again co-operation between owners will be the key to success and this is often difficult to achieve.

Human Demands on Feline Companions

Increasing feline populations both inside and outside the home undoubtedly influence the incidence and nature of behaviour problems in the domestic cat population, but another very important factor to consider is the effect of changing human expectations on the behaviour of this ideal modern pet.

One of the reasons for the cat's popularity is its relative independence and its ability to cater for its own needs, but while most owners are happy to know that their cat is getting on with its life while they are at work, they also expect their pet to provide companionship and social interaction for them when they get home. This desire to engage in low frequency but high intensity interaction with cats raises very specific challenges since the normal pattern of cat to cat interaction is one of very frequent but very low key communication. In other words cats will frequently pass the time of day with each other but do not engage in periods of intense communication, whilst owners spend many hours ignoring their cats but want to cuddle and pet them intensely when they return home. Such a fundamental difference in approach to social interaction is bound to bring tension, and education of cat owners in the unavoidable restrictions of natural cat behaviour is essential.

Certainly it is possible to educate cats to appreciate and value human interaction, and the process of early socialisation is important in preparing kittens for life in a human environment, but increasing feline acceptance of intense physical contact requires specific preparation and the manner in which young kittens are handled is critical. Remembering that the cat's primary defence strategy is flight, it is not difficult to understand why the acts of being picked up and restrained are potentially threatening, and periods of lifting, gently restraining and touching the kitten all over its body should therefore be routinely incorporated into any programme of early handling for kittens. Even when this has been done effectively it is important to respect the cat's natural behaviour and keep highly restrictive handling to a minimum, learning instead to respond to feline greeting behaviour and to use vocal interaction to enhance the relationship.

Pressures of a Domestic Environment

In addition to the effects that other cats and people have on the behaviour of the domestic cats there are also specific constraints of the domestic environment which lead to further compromise of their natural behaviours, and the successful integration of kittens into an average human household is something that requires a certain amount of preparation.

While genetic influences will help to determine how an individual reacts to novelty and challenge in its adult environment, adequate exposure to a wide range of stimuli during the early process of behavioural organisation will also be crucial in ensuring that the cat has a broad frame of reference with which to compare its later experiences. Kittens which benefit from an inherited boldness, a varied early environment and ongoing mental and physical stimulation in adulthood will undoubtedly benefit most from the diversity of the domestic environment and will be less likely to exhibit behaviours which compromise their life as a human companion.

Modifying the Human Environment to Cater for Feline Needs

In order to minimise the risks of feline behaviour problems it is essential to pay attention to the interactions between cats and between cats and people, but it is also important to modify the domestic environment in ways which will enhance its ability to cater for fundamental feline needs. Studying the nature of feline territories helps to highlight the qualities that cat's value and illustrates the importance of things like privacy, choice and hygiene in a feline context.

Providing Three-dimensional Space

The need for access to vertical space not only applies in multi-cat scenarios but also in single-cat households and provision of elevated hiding places is important in achieving the feline aim of minimising fear and anxiety within the home. Failure to cater for the natural feline defence strategies of flight and hiding can result in cats feeling threatened and increase the risk of developing problems related to aggressive behaviour, but it can also result in chronic feline stress which is manifested in inappropriate self-appeasement behaviours, such as over-grooming and over-eating. Simple alterations to the home in terms of providing radiator cradles and pieces of bedding on tops of wardrobes and kitchen cupboards can make a considerable difference to the quality of the environment from a feline perspective and in situations where anxiety is causing obvious behavioural changes the addition of a pheromone diffuser (Feliway®) to the home will also be beneficial.

Changing to Natural Feeding Patterns

The provision of immediate and unhindered access to

resources is an important feature of feline existence and is equally important in both single- and multiple-feline households.

In order to achieve this in relation to food it is advisable to adopt a policy of *ad libitum* feeding rather than stick to a rigid system of two meals a day, and this also makes more sense in relation to natural feline feeding routines as well as feline anatomy and physiology. After all, hunting behaviour is a time-consuming activity in the wild and cats undertake between 100 and 150 hunting attacks per day over a period of six to eight hours. These attacks lead to relatively small meals and interestingly the success rate of feline hunting behaviour is surprisingly low with only 10 per cent of them actually resulting in the acquisition of consumable prey. Consequently the ratio of energy expenditure to energy consumption is high and this leads to a very effective level of natural weight control. However, in the domestic environment when food is provided without any need for feline effort the ratio of energy input and output is likely to be altered and problems of unsuccessful weight control may result. In addition the tendency to provide meals on a twice-daily basis over faces the cat's relatively small digestive tract and results in cats failing to eat all their food in one sitting. Such behaviour is readily interpreted by caring owners as a sign that the cat is not happy with the food and selection of more palatable and more energy-dense diets is often the result. Not only does this increase the tendency for cats to hold out for a more palatable option but it also further upsets the balance between energy input and output and runs a very real risk of inducing problems of feline obesity.

Increasing the Effort to Acquire Food

In order to deal with problems of feline weight control attention needs to be given to both sides of the input/output equation and alterations in the methods of feeding need to be combined with changes in the provision of both mental and physical exercise for cats. The use of multiple-feeding stations around the home will encourage the cat to search for food and if relatively obscure locations can be added from time to time this will further increase the challenge.

So-called puzzle feeders can be used to increase the effort that is needed to gain access to the food source and such devices can significantly alter the amount of time and effort that the cat has to spend in the process of consuming its daily diet. Commercial puzzle feeders are available in some pet shops but it is easy for owners to make their own from a small plastic drinks bottle, in which they make small holes, just big enough to release the pieces of dry cat food, which are placed inside. As the cat knocks the bottle around the floor it will be rewarded with the arrival of the pieces of food and the fact that the bottle is transparent and makes a noise as it moves will help to keep the cat's interest. However, cats do need to learn to use these devices and in the early stages the holes in the bottle should be so large that the food falls out with minimal feline effort.

Providing Adequate Mental and Physical Exercise

Ensuring that cats expend sufficient mental and physical energy during the day is not only important in terms of weight control but also in terms of maintaining physical and mental fitness. Cats are designed to engage in short bursts of energy-consuming activity, which is often related to predatory behaviour, and intersperse these with significant periods of rest and relaxation. Provision of outlets for this energy-rich activity is therefore an important part of cat ownership and an understanding that predatory behaviour is stimulated in isolation of the sensation of hunger helps to explain the very real need that cats have for periods of predatory-motivated play.

The fact that predation is most likely to occur at times of dawn and dusk means that much of a cat's high-intensity activity will naturally occur at these times and failure to make specific provision for predatory responses during the day can result in cats whose night time and early morning activity becomes seriously disruptive to the domestic environment. In addition, a lack of opportunity to engage in predatory responses during play can lead to inappropriate predatory behaviour toward human 'prey' and owners can find their own ankles and hands being targeted.

Management of these behaviours is achieved by providing appropriate play on a regular and frequent basis and directing such natural responses on to appropriate toys. Fishing-rod toys, which stimulate feline interest through the features of rapid unpredictable movement and high-pitched noise are ideal for this purpose. Other items such as laser pointers can also be used to mimic the activity of potential prey, but it is important to remember that part of the predatory sequence is the act of catching and killing. Therefore the use of laser pointers, which can never be caught, as the only source of predatory activity is not appropriate and can even lead to the induction of compulsive behaviours which can be difficult to treat.

Catering for Feline Toileting Preferences

Hygiene is one of the important features of a natural feline territory and one of the most positive aspects of cat behaviour for many owners is their fastidious approach to toileting. The fact that they are already house trained when they arrive in their new home at just a few weeks of age is something that people find very appealing. However, this reputation for cleanliness leads to high expectations, and it is important to recognise that provision of appropriate toileting facilities is a prerequisite for acceptable and appropriate toileting behaviour.

When litter facilities are going to be provided so that cats can urinate and defecate within the boundaries of the home, it is important to ensure that the location provides all the qualities that a cat will search for in a latrine in the great outdoors, and attention to the provision of adequate privacy is essential. Cats will naturally eliminate at the periphery of their territory, away from other resources such as feeding stations and resting areas, and select locations where they will be undisturbed. Indoor latrines must therefore also offer seclusion, and sites which make the cat vulnerable or force proximity to other resources must be avoided. The fact that cats will often separate the functions of urination and defecation into different latrine sites when outdoors may lead to a need for separate litter facilities within the home and the provision of multiple indoor latrines which offer the cat choice will usually be beneficial.

In addition to considering the location of litter facilities it is also important to pay attention to the substrate that is offered within the tray and once again important lessons can be learned by observing the sort of material that cats naturally elect to eliminate in. In general terms preferred outdoor locations offer a soft and rakeable substrate, which can be easily manipulated with the paws during the process of digging an elimination pit and raking over the deposits, and therefore artificial litters should also be selected for these properties. Materials that are hard underfoot or are difficult to rake should generally be avoided since they are likely to decrease the cat's desire to use the tray and may even lead to problems of inappropriate elimination in other locations around the home. Adequate depth of litter must also be considered since cats have an innate desire to bury their excreta and shallow litter may lead to disruption of this important part of the toileting sequence. Likewise litters which are strongly scented are likely to interfere with the cat's natural instinct to bury until the scent of the deposit has reached an acceptable level, and use of these products can be a factor in initiating problems of inappropriate indoor elimination.

Ensuring Security within the Territory

Cats are territorial creatures and marking behaviour is an important tool in ensuring that unnecessary encounters with unfamiliar cats are avoided and territory is kept safe from potential invasion. Both urine and faeces can be used as a form of communication in this context, but the deposition of urine as a marker is by far the more common approach and the cat adopts a very characteristic stance when engaging in this behaviour. As a result of using this posture the cat ensures that the urine deposit is ideally positioned to be read by passing felines and that its message reaches its intended target. Such behaviour is perfectly normal and acceptable in the context of the great outdoors and potential encounters with other cats in the neighbourhood are sufficient justification for depositing some clear signals of occupancy around the boundary of the territory. However, marking behaviour is naturally restricted to the home and hunting ranges and whilst keeping the marks fresh is obviously important, in order to signal ongoing occupancy, the number of deposits made by cats which are confident in their right to occupy the location are likely to be minimal.

Marking within the core territory is not considered necessary and other behaviours such as eating, sleeping and playing are used to identify that area as a place of security and relaxation. An understanding of this use of marking signals is helpful when trying to interpret the inappropriate behaviour of indoor urine-marking, and by paying attention to the activities that identify an area as core territory, it is possible to increase feline confidence and security within the home, thereby decreasing the perception that marking is necessary in that context. Obviously identification of the reason for the cat's low level of confidence is going to be essential when managing problems of this nature but when considering prevention of urine-marking problems the emphasis is on labelling the house as a place of security, and thereby removing the need to mark it. Provision of adequate resources which are readily accessible and of privacy and choice in the home environment will therefore be crucial elements of the prevention programme and the use of pheromone signals (Feliway®) which confirm the familiarity of the location will also be beneficial.

Conclusions

The increasing popularity of the cat as a companion animal has led to a significant alteration in the role that it is expected to fulfil within human society and, whilst the majority of cats adapt well to the challenge, there are others which find the constraints of domestic life difficult to adjust to. Minimising the discrepancy between feline needs and human expectations is therefore essential if cats are to be relaxed in their companion animal role, and interpreting situations from a feline perspective will be a significant factor in achieving this goal.

9
Behaviour Problems in the Domestic Rabbit

E. Anne McBride, Emma Magnus, Georgie Hearne

Introduction

The third most popular mammal pet species in the UK is the rabbit; a species that has been kept has a pet for over 200 years, though it has been kept as a domesticated species for some 2,000 years. Attitudes to this animal have changed quite dramatically since the 1980s as demonstrated by advances in veterinary care, and more latterly pet insurance, in response to demand by owners. Since the 1990s there has been an increase in the house rabbit population and more recently an increased awareness of the role of the pet behaviourist when owners have problems with their rabbit's behaviour. Another reflection of the attitude change is the increasing number of rabbits in rescue societies estimated at 33,000 in 2002, and the increasing number of societies devoted to this single species, currently 300 in the UK alone. Whereas in the past rabbits with behaviour problems would have had a limited lifespan, either being relegated to low level care until they died, through being abandoned or euthanased.

The rabbit is a common but relatively unknown exotic species. Even simple but fundamental differences of herbivore versus carnivore are not always recognised, with owners providing lamb and rice diets to pet rabbits with diarrhoea as they would their pet dog/cat. Whilst meant with the best intentions, sadly such a lack of knowledge of species can lead to welfare issues both of a physical and psychological nature – as in the case above which did not survive this inappropriate treatment.

What Is a Rabbit?

Perhaps of all the common pet species, the rabbit is the most biologically unique. It is not a rodent, as are gerbils, chinchillas and mice. The European rabbit is the sole domesticated member of the Lagomorph family, which includes hares and jackrabbits. Yet even within this family the domesticated species has interesting and significant characteristics of its own, particularly in relation to rearing its young.

The rabbit is a herbivore which evolved in the harsh conditions of the Iberian Peninsula where the herbage is of low quality. It is a selective feeder which spends around 70 per cent of its active time eating. It has an efficient digestive system that includes the production of two types of faeces – soft ceacal faecal balls and hard faecal droppings. The former are caught from the anus and redigested in a process known as refection. This proper functioning of the gut relies on a high-fibre diet. The teeth grow continuously throughout the rabbit's life. They are designed to slice and grind thin strips of vegetation, such as grass and hay. Unlike those of rodents, they are not designed to gnaw hard foodstuffs. The provision of such feeds can lead to the misalignment of teeth, formation of sharp spikes, abscesses and subsequent death. The need for high-fibre content provided in thin strips which take up a substantial portion of the animal's day to eat is why it is strongly recommended that hay and fresh herbage forms the bulk of the diet, and commercial pellets are provided only as a supplement.

In natural circumstances rabbits are rarely seen to drink. This has led to a misconception that rabbits do not need to have a constantly available water supply. However, rabbits like all animals will dehydrate if they get hot or are unable to take in sufficient liquid. In nature this is less likely to happen: rabbits remain cool by going below ground or into the shade. In addition, by feeding primarily at dusk and dawn they obtain a lot of moisture through the dew that has settled on the herbage. It is important to ensure that rabbits always have a freely available source of water.

The rabbit is a prey species, indeed it is one of THE prey species contributing some 20 per cent of the diet to almost 30 different predator species in its native habitat. The rabbit's size means that it provides a neatly packaged, substantial size meal. It has enemies that hunt it at ground level such as fox, cat and dog; other predators attack below ground, including these three, but also badger, stoats, weasel, polecat and their domesticated cousin, ferrets. For birds of prey, the rabbit provides a welcome addition to the diet, and young nestling buzzards are fed almost exclusively on rabbits. Of course the predator which allies itself with those that can fly, run or dig and thus catch rabbits is man with his falcons, running dogs, terriers and ferrets, as well as pet cats. Having evolved as Mother Nature's fast-food option, it is not surprising that much of its behaviour is orientated towards detection and avoidance of predators.

Rabbits live predominately underground in a complex system of tunnels known as a warren. They spend most of their life in the dark below ground, or semi-dark when they emerge between dusk and dawn. This means they are less visible to predators and this also has relevance when rabbit communication signals are considered. One aspect of this is that rabbits do not show obvious signals of pain. This may have evolved because animals that show weakness are likely to be the preferred target of a predator.

The rabbit has acute senses of sight, hearing and smell to enable it to detect potential danger. By living in groups, the efficiency of these 'radar' is increased, often with those on 'sentry' duty choosing higher places such as molehills to stand on. The first response of an individual to danger is to freeze, the second to flee. Only if caught will a rabbit defend itself by kicking and biting. A rabbit will also warn others nearby of danger by loudly thumping its feet on the ground, a noise that travels both above and below ground.

The need to be able to flee to a place of safety is so important to rabbits that they alter their feeding – vigilance behaviour depending on how far they are from cover and how many other rabbits there are around them. Lack of provision of raised sentinel places and cover in the form of pipes and boxes in rabbit pens can result in stress-related problems including agoraphobia and aggression.

Rabbits live in groups the size of which is dependent on population density and availability of resources including suitable sites for burrowing and soil stability – the more stable the soil the deeper the tunnels that can be constructed and thus house more animals. Where burrowing is easy the normal group comprises a single female and her mate. Larger groups tend to have more females than males. Groups tend to remain stable during the breeding season (January – August in the UK) with a strong hierarchical structure. This breaks down in the remainder of the year, allowing new individuals to integrate into groups. In general females stay with their natal group and juvenile males migrate to new groups, often having spent several months as satellite individuals whilst the breeding season continues. Male hierarchies are linear during the breeding season which controls access to breeding females and within the group the males co-operate in defence of territorial boundaries, through scent deposition and by deterring intruders. Whilst a dominant female will emerge they tend to have a looser hierarchical structure. This may be due to the fact that they are more closely related. Having said this, females will become extremely aggressive over resources, in particular nesting sites, during the breeding season and fights to the death can occur. This sudden and dramatic increase in aggression in early spring is often reported by owners whose animals have lived together peaceably for several months, even the best part of a year.

Rabbits display aggression through an escalating series of signals that are often not easily recognised. As with all species this can be motivated by different emotions.

Competitive Aggression

This often arises between rabbits of the same sex, and less commonly in a domestic situation, of opposite sexes.

As we have discussed, male rabbits are particularly territorial during the breeding season, when they form linear hierarchies within their groups. In an encounter with a trespassing individual, the highest status male will often approach the unknown individual, stopping to nibble restlessly at grass, sometimes engage in some marking behaviours or frantically dig at the ground. If the trespasser does not retreat a fight is likely to ensue. The behaviours likely to be seen before outright aggression include chasing, scraping of the ground and stiff-legged runs past each other. Physical violence rarely occurs at this stage; usually these rituals forestall it. When it does occur it consists of powerful kicks with the hind legs with rabbits often grabbing each other's necks as they kick. Competitive aggression between females usually follows a similar pattern. However, it can be more serious.

Defensive Aggression

Aggression of course can be motivated by fear of a threat, commonly known as fear-related or defensive aggression. This is a common cause of aggression towards people and again may result from the rabbit's early warning signs having been misunderstood.

When the rabbit becomes frightened the behaviour displayed will usually begin with subtle signals, such as an increase in muscle tension as the animal freezes, its ears will be held back flat to the head and sometimes they will growl and may try to flee. If these signals are ignored the rabbit may well shift its weight on to its back legs, which can give it an appearance of being slightly hunched. Lunging forward and biting often swiftly follows this. This may be a bite and release as the rabbit runs away, or it may hold and kick.

Other Social Behaviours

Given that rabbits spend lots of time in the dark and also try to keep their precise location at any one time hidden they don't display much in the way of visual or vocal communication. This causes difficulties for humans who are used to living with species that communicate predominantly through these methods, such as dogs and cats. They do however communicate socially, although this too is subtle. Affiliative signals include tooth grinding, mutual grooming and seeking close contact.

Scent is the main method of communication that rabbits will use, and this originates in three different glands on the rabbit's body:

- Sub-mandibular gland. This is located under the chin, and domestic rabbits will often be seen rubbing their chin on objects, people and other rabbits. This behaviour acts to pass a common scent profile to members of the group and objects within the territorial boundaries and so may well act as a territorial marker.
- Anal-gland secretions. These are deposited with the hard droppings. Rabbits tend to use latrine sites for elimination and all members of the group use these. They are often located on higher ground such as molehills and tree stumps where they act as both a visual marker of territorial boundaries from where the scent can be wafted further afield. This behaviour has been used to advantage by the pet owner who can toilet train their rabbit to use litter trays or single areas in the garden as latrine sites. It should be noted of course that rabbits also drop faeces as they forage and whilst the bulk might end up in the litter tray, others will be spread around the environment.
- Inguinal gland secretions. These are deposited with urine, in particular during courtship, and sometimes during territorial disputes. Spraying consists of the animal running past the target at speed, twisting their hind legs and accurately spraying a jet of urine.

Development

Like puppies and kittens, rabbits are born in an undeveloped state: they are naked, blind and deaf and barely able to move. However, this is where the similarities end. Rabbit maternal care is radically different from that of other species that have undeveloped young.

Doe rabbits give birth after a pregnancy of some 28–30 days. A fur-lined grass nest is located at the base of a shallow burrow, the entrance of which is blocked with earth whenever the mother is not present – which is most of the time. Rabbit mothers only visit their young once every 20–24 hours and then for a mere three minutes to suckle them. The young are capable of keeping themselves warm and dry by eliminating spontaneously once they have fed and then burrowing down into the nest material, only emerging again shortly before the doe is due to return.

Young rabbits will begin to emerge from the burrow in their third week and will be weaned at the end of their fourth. It is then that they start to make their way in rabbit society. The period from birth to sexual maturity is a mere three months (this is delayed in some of the larger domesticated breeds). However, in the wild a female born in the early part of the season is likely to have had one or two litters before the season ends.

It may be considered by some that such a brief time with the parent and short period of development would mean that most of rabbit behaviour comprises innate fixed action patterns, and more complex developmental issues such as socialising and other behavioural organisation are not relevant, or at least not important. This is not the case.

Studies have shown that handling rabbits between ten and twenty days old can have a profound impact on their later willingness to approach humans. In addition, exposure to other species at this age, including potential predators, can reduce a rabbit's reactivity when they encounter these individuals later on. In one experiment, rabbit kittens which had received early exposure to cats did not react fearfully when a cat approached them later on. A similar group, handled only by humans showed an increased likelihood to approach humans at weaning age, while non-handled controls avoided both cats and humans, suggesting that rabbits have a socialisation process.

Effects of early handling seem long lasting, as, in one experiment, handled females raised to adulthood were superior to non-handled individuals in breeding performance. Perhaps one explanation is that early handling reduces stress on the doe, enabling her to raise a litter.

Further research to pinpoint this socialisation period and its profile is required. However, the evidence to date strongly suggests that rabbits need to be handled early to prevent later problems. Early handling of kittens should be carried out carefully, ensuring that the hands of the handler are first anointed with the scent of the doe.

Owners need to remember that a rabbit which has been well socialised, and thus has the ability to accept the relevant species, will still remain reactive to certain situations. This is because, fundamentally, they are a prey and flight species and anything can be perceived as a threat and cause anxiety and fear.

Being reliant on flight and escaping to a safe place in a natural environment that is continually changing means that rabbits have to have a constantly updated knowledge of the area in which they live. This in turn means that rabbits have a high motivation for exploration and an ability to learn and remember spatial maps. Rabbits are able to learn through habituation, classical and operant conditioning, as well as by observing the behaviour of others. The principles of training can be applied to rabbits, which can be taught a variety of tricks and cues such as coming when called, fetch and even agility. Indeed rabbit agility is gaining popularity as a sport.

Preventive Measures

Prevention starts with knowledge. Prospective owners need to decide if a rabbit is the suitable pet for their lifestyle and to consider which breed is most appropriate. The acquisition of a rabbit should be taken no more lightly than that of a cat or dog. Rabbits should not be considered a 'short-term', 'convenient', 'low maintenance' pet. When kept in appropriate conditions some rabbits can live for eight years or more, though for giant breeds this figure may be nearer five.

Rabbits come in a variety of breeds and sizes, from the tiny dwarf breeds that weigh around a kilogram to the giants which can reach the size of a corgi and weigh 10 kilograms. Different breeds also have different temperament characteristics, with the smaller breeds generally being more reactive than the larger. This in part reflects the functions for which they were originally bred, the larger breeds, having been used for meat or fur production or as laboratory animals, were selected for a less reactive more phlegmatic character. Likewise the lop breeds also tend to be less reactive than similar sized more rangy breeds such as Netherland, Dutch, Polish or, somewhat misleadingly named, Belgian Hare. Unfortunately, many rabbits are purchased on impulse from pet shops, newspaper adverts or at local breed shows. They are often bought as a child's pet – a pet that often does not maintain the child's interest long term and may then be less than adequately cared for – resulting in physical and behavioural problems.

Having decided on a rabbit and breed, gender is the next consideration. Ideally two rabbits – one of each sex – should be purchased. They should be of similar size and breed, the easiest option is a litter brother and sister. In this way the rabbits will have constant companionship in a natural group, with none of the possible problems associated with introducing strange rabbits to each other at a later date. As soon as possible both members of the pair should be neutered, reducing the likelihood of any seasonal aggression and unwanted youngsters.

Housing is obviously a major consideration and to some extent will be dependent on an owner's preference. Some will wish to have their rabbit as an indoor pet, as one might have an indoor cat. Others will prefer to keep their rabbits outside, preferably in an integrated hutch and run complex. Whichever arrangement is chosen there are some basic principles that need to be adhered to in order to ensure the welfare of the rabbit and continuation of a good human-rabbit relationship.

Rabbits must be provided with sufficient space so that they can move freely, stretch to full length, hop, jump and even run. They should be provided with places in which they can hide, take shelter from the sun and rain, stand up on and be away from draughts. Rabbits need to be kept in a well ventilated environment though being exposed to draughts can result in respiratory problems. Likewise, whilst rabbits will enjoy some sunbathing, they can easily become overheated and must have access to cool, shady areas. After all they are wearing a fur coat that has often provided good insulation to human beings.

Rabbits are curious creatures and will play with objects in their environment and should be provided with suitable toys, such as the cardboard inner tubes of toilet rolls, toddler teething rings and small footballs. The provision of food should also be adapted to encourage the rabbit to graze. This can be achieved by providing hay in a rack, or within a brown untreated paper bag with some dried or fresh herbs, or threading pieces of green vegetables on to covered wire which can then be suspended from the roof of the hutch or indoor cage. Small branches of untreated fruit trees and leaves can also provide very attractive occupational therapy. There are several toys for rabbits on the market, but not all are totally suitable, such as those that encourage the rabbits to use their teeth inappropriately for gnawing.

Last but not least is an appropriate diet comprising mostly hay and fresh greens, with a limited supply of commercial feed and continual access to fresh water. In this way the rabbit will be occupied in a natural manner, namely foraging and grazing, and the likelihood of dental problems or disease associated with obesity are minimised.

Behaviour Problems

It is only recently that rabbit owners have considered seeking professional treatment for behaviour problems and this means that the area is in its infancy.

The table below outlines four typical rabbit behaviour problems and some of the many possible causes. This highlights the importance of veterinary referral and the need for an extensive case history. Behaviour problems may be multi-causal and treatment rationales will vary from case to case.

Possible Causes	Aggression to other rabbits	Aggression to people	Spraying/ inappropriate elimination (please note there there is a difference)	Pica (eating non-nutri tional items)
Pain	x	x	x	x
Medical issue	x	x	x	x
Poor socialisation and other behavioural organisation	x	x		x
Inappropriate handling		x		
Hormonal influences	x	x	x	x
Learned response	x	x	x	x
Anxiety response	x	x	x	x
Redirected	x	x		
Territoriality	x	x	x	

To illustrate the complexity of rabbit behaviour problems two case histories and possible differential diagnoses are described in brief below.

Muffin
Pet's name: Muffin
Breed: Mini-lop
Sex: Male – entire
Age obtained: Nine weeks of age
Previous environment: Outside hutches

Current environment:	House rabbit
Age:	six months
Medical history:	Physical examination was unremarkable.
Diet:	Hay, fresh grass and household greens, as well as one quarter of a rich tea biscuit each evening
Month of consultation:	April

Muffin was obtained at nine weeks of age from a breeder, being a house-rabbit from this age onward. He was litter trained relatively easily, although he will drop pellets around the house occasionally. His daily routine depends on whether there are people at home, as he is kept in an indoor hutch rather than left in the house at these times. His owner feeds, grooms and gives Muffin his freedom from 5.30p.m. onwards each day, after returning home from work.

The client has sought advice because Muffin urinates on her and 'no one else'. He will also urinate and defecate on the sofa, and on the client's bed. She has also seen him spray two to three times on the side of the sofa after being interrupted from urinating in his 'usual spot'. In the presence of the owner he will usually visit the site two to three times before eliminating. The owner feels that if she is not in the room he will do it immediately. (N.B.: whether this is true is not known, the client leaves the room and on returning there is rabbit urine on the furniture. His behaviour preceding the event is unknown).

The most recent example of Muffin's behaviour was during three hours of free time when the owner's male flatmate was at home. The rabbit went into her room for approximately twenty minutes and urinated and defecated on the bed. The previous incident occurred when the owner had Muffin on her lap. She had just changed her clothes as her mother's rabbit had just urinated on her and Muffin climbed on her chest and 'seemed to be searching for more rich tea biscuits'. He then urinated on her. In these situations the owner's response is to shut Muffin away for fifteen minutes.

Although Muffin urinated on his owner on the second day she brought him home (at the beginning of December), it stopped around Christmas. Muffin's behaviour started again in February, the most recent events were during April.

DIFFERENTIAL DIAGNOSES

It is possible that the lack of social interaction, or company in a gregarious species, particularly when there are unfamiliar scents entering the territory could cause Muffin a degree of stress. As a prey species rabbits are particularly susceptible to the effects of adrenal hormones. As well as physiological illness, behavioural effects of stress include anxiety that can cause over-vigilance and 'over-reaction' to fear-eliciting stimuli. Pain, unfamiliar surroundings, loud noise and proximity of predators can all cause stress responses in rabbits. Urination and defecation can both be behaviours that are seen during the 'flight or fight' response and this was considered as a possible cause of Muffin's behaviour.

Physical examination prior to referral would have been expected to rule out many causes of pain including dental disease, spinal fractures or osteoporosis. As his surroundings have remained relatively stable, the firework 'season' has finished and Muffin lives as a house-rabbit in a flat with no other pets, it is unlikely that his environment is causing him unreasonable anxiety. As Muffin can be held by the owner's flatmate, and seems only to direct his toileting at objects that are associated with the owner herself, anxiety or fear of particular stimuli was considered to be unlikely.

DIAGNOSIS

It is likely that Muffin is displaying normal behaviour in an inappropriate context. It seems to be related to either territorial marking behaviour or possibly an over-attachment or sexual bond to his owner. It is also likely that elements of both could contribute to this problem.

Sooty and Fluffy

Pets' names:	Sooty and Fluffy
Species:	Rabbits
Breed:	Dwarf cross
Sex:	Female – entire, Female neutered respectively
Age obtained:	nine weeks (approximately)
Previous environment:	Pet shop
Current environment:	Hutch and Run with indoor access each evening
Age:	Six months and two years respectively
Medical history:	Routine vaccinations and spay for Fluffy
Diet:	Commercial mix and fresh vegetables
Month of consultation:	March

Fluffy was obtained about two years ago, along with her sister Dotty. Both rabbits lived in the same hutch and cohabited without any incidents. Unfortunately Dotty died unexpectedly at eighteen months of age and the family wanted to obtain another companion for Fluffy.

Sooty was bought from a pet shop as a 'six-week-old male rabbit' and the initial veterinary check soon after he was purchased confirmed this. From the beginning the family kept Sooty in a separate hutch and worried that Fluffy was 'intolerant' of Sooty. They attempted to introduce the rabbits on neutral territory; this was the concrete area outside Fluffy's hutch. Each time Fluffy would chase Sooty around the pen, apart from this the children felt that when together the rabbits interacted through a mutual tolerance of each other's company. In the preceding months Fluffy's chasing has become more frequent and Sooty has begun to defend herself aggressively. They will now fight on sight of each other. The children are very anxious about the behaviour but keen for the rabbits to interact. Observation found that they would cry out and pull the harnessed rabbits apart if there was any interaction between them at all, anticipating the aggression.

Both daughters sometimes have trouble catching and handling the rabbits and were keen to improve their relationship with their pets.

Accurate observation and sexing established that Sooty was in fact female and if put in the concrete area adjacent to the hutches the rabbits would soon begin to interact, Fluffy usually chasing Sooty. Both rabbits were on leads and harnesses and were separated at this point. When in a grassy run they were less interested in each other and grazed for around five minutes before passing each other in their feeding pathway. They were able to be distracted from each other with dandelion and strips of carrot.

DIFFERENTIAL DIAGNOSES

As in other species, pain can result in aggressive behaviour. Dental disease or the formation of sharp hooks on the molars can be extremely painful and rabbits are also susceptible to developing painful musculo-skeletal disorders such as arthritis or vertebral spondylitis. Whilst both rabbits had been referred as having no medical problems, further medical work would be appropriate if the situation continued to deteriorate, or if the rabbits start to show physical signs of illness.

A learned pain association may have developed if either rabbit was experiencing musculo-skeletal pain and was pulled away from the other rabbit whilst wearing the harness. The children's anxiety led them to pull the rabbits quickly away at the same time as they were investigating one another, which may have caused the rabbits anxiety through a conditioned fear response leading to aggression in future. If pain of this type was causing or contributing to the behaviour, it could be that the rabbits were being aggressive when handled or when anticipating handling in the hutch. At the time of consultation they were both able to be picked up, groomed, handled and harnessed without any defensive aggression or avoidance.

It is also possible that limited previous exposure to open spaces or nearby predators has caused fear of novel stimuli in one or both of the rabbits. This type of fear reaction undoubtedly causes a degree of stress that may lead to over-vigilance or a decreased threshold for aggression. As both rabbits were kept in a covered area adjacent to the house it was unlikely that nearby cats or birds of prey in overhead trees were causing their behaviour. Similarly both rabbits would graze outside in a run separately and Fluffy spent much time in the area when she was housed with her sibling.

DIAGNOSIS

Most aggressive encounters in rabbits are intra-sexual with females tending to be the more aggressive sex. Sooty's gender was established wrongly at the end of the breeding season. This is a common mistake as at this time the testes regress into the abdominal cavity. Sooty was purchased in October, her aggressive responses to Fluffy were noticeable in around the beginning of February and have increased in frequency until the time of consultation, suggesting the aggression is due to intra-sexual competition related to the breeding season.

Conclusion

As with all companion animals, the treatment of behaviour problems is not available in a recipe-style format. This is particularly relevant in the field of rabbit behaviour, which is relatively new and the development of our understanding of this species and appropriate methods for behaviour modification are progressing rapidly. All behaviour practitioners, whatever the species they are treating, need a sound knowledge of its ethology, learning theory and its husbandry requirements to complement practical experience. This is also true for the rabbit, but people often consider that these are 'easy to treat' animals, or can be treated using similar methods which have gained acceptance in the field of dog and cat behaviour modification. This is not the case and can lead to exacerbation of existing problems or the development of new ones.

10
Rage Syndrome: In our dogs' minds?

Anthony L. Podberscek

Aggression in dogs is one of the most common problems treated by behaviour consultants, and can be categorised in a number of ways, for example, dominance-related, predatory, territorial, possessive and interdog aggression. One type of aggression which is difficult to categorise is rage syndrome. The difficulty arises from our having little knowledge of what actually causes it. To this end, it is usually placed in a category of aggression known as 'idiopathic aggression', which basically means aggression for which there is no known or understood cause. In this chapter we shall look at the current state of knowledge on this perplexing form of aggression.

What Is It?

Rage syndrome refers to sudden, unpredictable, unprovoked and uncontrollable violent aggression towards a person, other animal or object. The term became popular during the early 1980s, when there was a series of newspaper articles, e.g., 'Top Dogs Turn Nasty', *The Sunday Times*, 1981, and magazine articles – 'Solid Cockers Condemned', *Dog World*, 1981). Those affected were of solid coat colour (such as red/golden and black on aggression in English Cocker Spaniels). These articles were in response to a scientific paper written by animal behaviourist Roger Mugford on the types of aggression in English Cocker Spaniels that he treated at his clinic. Indeed, the publicity was such that rage syndrome became synonymous with this breed, leading some to call it 'cocker rage syndrome'. And this association was further enhanced in the media, including a feature article (along with cover photo) in *Dogs Today* in October 1992: 'Cocker Rage Syndrome: Madness or Make-Believe?' People have even suggested that there are references to a Cocker 'raging' in nineteenth-century England: that Elizabeth Barrett-Browning, in her letters to Robert Browning, makes references to rage-like behaviour in her male, golden Cocker Spaniel, 'Flush'. So I asked scholar Maureen Adams, who has studied extensively the works and letters of Elizabeth Barrett-Browning, if she knew of any such references to Flush's behaviour. She provided me with the following excerpts from Elizabeth Barrett-Browning's letters:

> Critics who bark the loudest commonly bark at their own shadow in the glass, as my Flush used to do long & loud before he gained experience & learnt the [in Greek] 'Know Thyself' in the apparition of the brown dog with glittering dilating eyes (8 August 8, 1845)

> 'For Flush, though he began by shivering with rage & barking & howling & gnashing his teeth at the brown dog in the glass, has learnt by experience what that image means, . . . & now contemplates it, serene in natural philosophy' (24 March 1846)

Also, Maureen found that Flush did bite Robert Browning on two occasions.

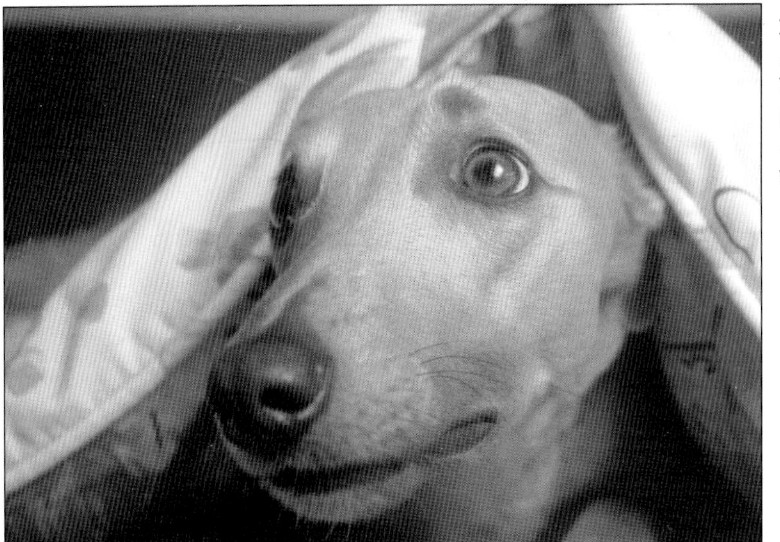

Photo: David Appleby

9. Fear of loud noises may cause a dog to seek a location that improves its emotional homeostasis and capacity to cope

Photo: David Appleby

10. Animals may resort to the use of aggession as a means of controlling their environment and regaining their emotional homeostasis

Photo: Jon Bowen

11. Urine spraying – a natural method of communication for the cat, but unpleasant for humans if performed indoors

Photo: David Appleby

12. Some multi-cat households are made up of sibling pairs such as these, but in many cases there is no such basis to the feline community and cats are expected to live with total feline strangers

Photo: Simona Tigwell

13. Rabbits are curious creatures and will play with and investigate objects in their environment

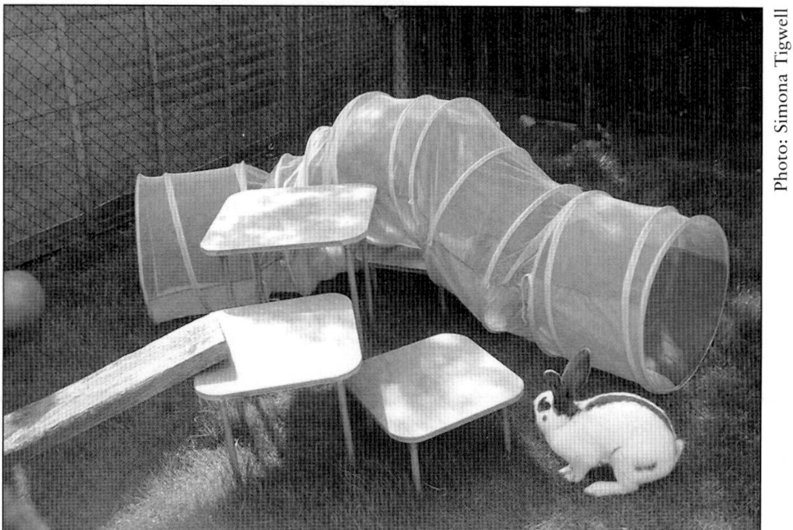

Photo: Simona Tigwell

14. Garden and indoor toys can provide security, shade and stimulation

Photo: Simona Tigwell

15. The rabbit's natural behaviour makes it possible to use a litter tray

16a and b.
Creative thinking can provide environmental stimulation for dogs in a shelter environment

Photos: Julie Bedford

17. Environmental enrichment for the cat in a shelter should include different levels and places to hide

Photo: Julie Bedford

While these are interesting accounts of Flush's behaviour, it's unlikely that they are indicative of rage syndrome. Indeed, Maureen found that Flush was very protective of Elizabeth Barrett-Browning and believes that this may have been why he bit Robert Browning.

This is not to say, though, that Cockers do not suffer from rage syndrome, rather that the breed seems to have received undue attention in relation to it, and that the implication is that other breeds do not suffer from it. Fortunately, not all publicity for the English Cocker Spaniel has been bad, though. Scamper, the golden Cocker in Enid Blyton's *Secret Seven* books, had an impeccable temperament, apparently.

What Happens?

From the outset, it is very important to stress that the information provided here is based on the limited amount of information available on the topic, and that in many cases a diagnosis of 'rage syndrome' has been based purely on behavioural signs alone.

Owners report that their dogs are normally healthy and good-natured, but that from time to time, and without any warning, the animal will become violently aggressive (biting – barking may occur, as well), usually towards the owner or other member of the household, although attacks towards another dog or an object have been recorded in the some cases. During the attack the dog's eyes appear glazed and its pupils are dilated, and some people have reported that the eyes appear red in colour. In addition, some dogs drool, vomit, defecate and/or urinate during an attack. By all accounts, the dog does not seem to know what it is doing and its behaviour is uncontrollable. Attacks towards people seem to be provoked by very simple and seemingly inconsequential interactions, for example, the owner bending down to pat the dog, the owner brushing past the animal, or the owner giving a command in a normal and friendly manner. Sometimes, just before an attack, owners have noted that their dog's behaviour changes: the dog appears to be in a trance – it stares or growls at nothing in particular. In other cases, attacks occur soon after the dog has just woken up. After an attack, which usually lasts for up to five minutes, the dog appears to be unaware of what has just happened, is tired/exhausted, and usually goes off to sleep for thirty minutes or more. Afterwards, the dog appears normal and friendly again.

This problem has been recorded in dogs as early as six months of age, but most reported cases involve dogs from one to three years of age. Male dogs are most commonly implicated, but it is not possible to say anything about the neuter status of these dogs, as it has not always been reported in the literature.

Interestingly, the dog in question often occupies high status in the owner's home (a lack of inhibition is present in the relationship with the owner leading to agonistic behaviour) and it may also display territorial aggression.

Breeds of Dog Involved

As mentioned earlier, rage syndrome has become synonymous with the English Cocker Spaniel. However, a number of other breeds have demonstrated this condition, or at least signs indicative of it. These include: American Cocker Spaniel, Bernese Mountain Dog, Bull Terrier, Chesapeake Bay Retriever, Dobermann, English Springer Spaniel, German Shepherd, Golden Retriever, Jack Russell Terrier, Lhaso Apso, Pyrenean Mountain

Dog, Rottweiler, St Bernard, West Highland White Terrier and Yorkshire Terrier. It is important to note here, though, that in most cases only one dog of each breed has actually been reported on in the scientific literature, and usually only behavioural signs have been used as evidence.

What Are the Potential Causes?

There are a few theories as to what might be causing these outbursts of sudden aggression in dogs.

STATUS-RELATED AGGRESSION

Although owners will claim that they did nothing to provoke the attack on them, behaviour consultants have noted that these people often have a problem with status-related aggression with their dog; that the dog sees itself as having a higher status in the household than the owner. In these cases, aggression can occur when the owner inadvertently does something that the dog doesn't like. These attacks will appear unprovoked because they are often simple, seemingly innocuous things, such as the owner going to pat the dog, to step over the dog, or to take something which is near the dog. Indeed, my research with James Serpell on 1,109 English Cocker Spaniels showed that those which showed aggression 'suddenly and without apparent reason' also often showed signs of status-related aggression (e.g., aggression when disciplined, aggression when reached for or handled).

It is also important to note that different breeds of dog, and indeed different individual dogs, will have different thresholds for aggression. So it is possible that a breed such as the Cocker Spaniel has a low threshold for aggression, and this may be partly why it has attracted more negative media attention than some other breeds of dog.

COMPLEX FOCAL SEIZURES: BEHAVIOURAL EPILEPSY

This is a type of epilepsy which manifests itself in the form of behavioural change, rather than the classic fits/convulsions we usually associate with epilepsy. Epilepsies can be broken down into two main groups: primary (idiopathic) and secondary. The primary epilepsies are those for which there is no identifiable brain abnormality other than the seizures themselves; the cause is unknown. Secondary epilepsies are those in which seizures are the consequence of an identifiable lesion (e.g., tumour) or other specific cause. Primary epilepsies can be further broken down into generalised seizures (involving both hemispheres of the brain) and focal seizures (abnormal activity in one region of a hemisphere of the brain). Generalised seizures include the ones most of us associate with epilepsy: loss of consciousness, convulsions and foaming at the mouth. Focal seizures can be either simple (consciousness is preserved during the seizure) or complex (consciousness is impaired during the seizure). We are interested in the complex ones, as this type of epilepsy can manifest itself as behavioural change (e.g., aggression) with impaired consciousness, general signs in common with rage syndrome. Other behavioural changes which have been observed in dogs during a complex focal seizure include sudden fearfulness, barking, 'fly-catching' (snapping at the air as if trying to catch flies) and tail chasing. In some cases, vomiting and diarrhoea have also been reported during the seizure. In cats, complex focal seizures manifest themselves in the form of confusion, defensiveness, aggression, hissing, moaning, and involuntary licking and chewing movements.

Focal seizures are thought to originate from a variety of places in the brain: the temporal lobes, limbic system, frontal lobes or the occipital lobes. If the location of the origin of the epileptic seizures is known, then a more specific diagnosis of the condition can be made, e.g., temporal lobe epilepsy.

When aggressive behaviour is a feature of an epileptic seizure, typical characteristics of seizures must accompany the aggressive act. That is, the seizure should begin suddenly and without provocation, last for a brief period (approximately one to three minutes) and end abruptly, with the violent act occurring in a context of impaired consciousness (dog is disorientated/ unresponsive).

And while most seizures appear to occur spontaneously, they may be precipitated by a variety of factors. For example, sleep deprivation, emotional stress and even heat in unspayed female dogs can precipitate a seizure. The most common type of trigger in people is flickering light, usually from a TV. Certain sounds can also act as a trigger.

However, although aggression during an epileptic seizure is possible, it has not been frequently documented in the medical or veterinary literature. And yet associations between epilepsy and violence prevail in the popular press and among scientists. A major problem is that some scientists and practitioners fail to distinguish between aggressive behaviours that are directly linked to seizure activity and aggressive behaviours that are related to other factors, such as the presence of other underlying brain abnormalities.

NONEPILEPTIC SEIZURES

These are seizures which appear to be epileptic in nature (e.g., they may appear to be a complex focal seizure), but in fact are not due to epilepsy. Instead, other types of brain dysfunction/disturbance are involved. In the medical literature, people demonstrating uncontrollable rage attacks, which are not due to complex focal seizures, are thought to suffer from intermittent explosive disorder (sometimes also referred to as episodic dyscontrol syndrome or rage). This is characterised by repeated acts of violent, aggressive behaviour in otherwise normal people, and the behaviour is out of proportion to the event that triggers or provokes it. Attacks are followed by fatigue, amnesia and sincere remorse. Sufferers are usually male and it is thought that problems in the temporal lobe/limbic system areas of the brain are the cause.

The problem we face initially with diagnosing what is actually causing the sudden aggressive outbursts in dogs is that the behavioural signs are common to all three causes detailed above. Without further investigation, we cannot make a definite diagnosis.

Diagnosing the Condition

The challenge is being able to diagnose the problem correctly. A full medical, behavioural and neurological examination is required. It is important to remember, too, that aggressive outbursts may be the result of something other than rage syndrome, for example, brain tumours, toxicity (e.g., lead poisoning), infectious/inflammatory diseases, head trauma and metabolic disorders (e.g., low blood sugar).

I believe that rage syndrome should only be used to

refer to *nonepileptic* behavioural seizures which manifest themselves in the form of sudden aggression. As I have said, in people, these types of seizures have been labelled intermittent explosive disorder, episodic dyscontrol syndrome or rage attacks. When we have evidence that the aggressive outbursts are actually epileptic seizures, then the diagnosis should be behavioural epilepsy/ complex focal seizure. But just to complicate things a little, human sufferers of nonepileptic behavioural seizures sometimes also have epilepsy. So we need to consider the possibility that a dog showing behavioural seizures may be suffering from both rage syndrome and epilepsy. Where there is no evidence of behavioural seizures (epileptic or nonepileptic), a diagnosis of status-related aggression would be more appropriate – this is not rage syndrome.

So how do we find clinical evidence of behavioural seizures? To date, the best diagnostic tool is an electroencephalogram (EEG). This is a recording of the electrical activity of the brain, abnormalities of which can indicate a variety of brain disorders, including epilepsy. Complex focal seizures always show an abnormal EEG *during* a seizure. For nonepileptic seizures, the EEG will always be normal during a seizure, while non-specific abnormal activity occurs *between* seizures.

How Common is the Problem?

To date, there have only been a handful of suspected cases which have been fully documented in the scientific literature. In addition, reports from both UK and USA behaviour consultants suggest that very few cases have been seen in clinics. From my research on English Cocker Spaniels, it was found that approximately 8 per cent of these dogs regularly showed aggression 'suddenly and without apparent reason'. An earlier Dutch study on 404 Bernese Mountain Dogs found that approximately 20 per cent of them showed 'intermittent attacks of blind aggressiveness towards their owners'. However, these were both questionnaire-based studies, and full behavioural and medical/neurological assessments on these dogs were not conducted. Consequently, we can't say anything about the dogs in terms of how many actually suffered from rage syndrome. Indeed, from my research on English Cockers, most of the potential ragers also showed behaviours indicative of status-related aggression.

What we do know is that seizures are the most common neurological problem reported in dogs: approximately 0.5 to 5.7 per cent of dogs have experienced seizures sometime in their lives. This is most likely an underestimate, though, as many people do not know that seizures may manifest themselves in the form of behavioural change and because owners may not bother to take their dog to their veterinarian if the seizures are minor or infrequent. A recent study on Poodles and Dalmatians who had seizures showed that the majority suffered complex focal seizures, although none manifested themselves in the form of sudden aggression. The behavioural manifestations observed were 'fly catching' and sudden fear.

Can It Be Inherited?

There certainly is potential for this because, based on pedigree analyses, a genetic basis for primary (idiopathic) epilepsy in several breeds has been shown. The breeds included are Beagle, Belgian Tervuren, Keeshond, Dachshund, German Shepherd Dog,

Labrador Retriever, Golden Retriever and Collie. As to whether complex focal seizures manifesting as aggressive behaviour are heritable, the jury is out, as not enough studies have been conducted.

From my own research on English Cocker Spaniels, it was clear that solid-coloured Cockers (red/golden, black) displayed most types of aggression, including 'suddenly and without apparent reason', much more commonly than did parti-coloured Cockers (e.g., blue roan, black & white). And because solid- and parti-coloured Cockers come from distinct bloodlines, this is a pointer to there being a genetic basis for the difference. However, this is not evidence for rage syndrome being genetic, as it was not possible to determine which, if any, of the dogs in the study actually suffered from it.

How to Treat the Condition

If a dog is found to be showing only status-related aggression – and therefore is not rage syndrome – a behaviour consultant can advise the owner on how to change their behaviour to bring about a more balanced and harmonious dog–owner relationship, with the owner firmly in control. The chances then of aggressive outbursts from the dog should be very greatly reduced or even extinguished.

For dogs suffering from epileptic and nonepileptic behavioural seizures, manifesting themselves as sudden aggression, anticonvulsant medications such as phenobarbitone, primidone and potassium bromide are the drugs of choice. Nonepileptic seizures may also be controlled using other mood-stabilising drugs, such as clomipramine. In addition, if specific triggers (certain sounds, owner behaviour) for these seizures can be identified, then it may be possible to advise the owner on how to avoid them.

Conclusion

At the end of the day, we know very little about rage syndrome. And yet, unfortunately, many people use the term too readily when diagnosing dogs which show unpredictable aggression, especially if that dog happens to be an English Cocker Spaniel. Without conducting full medical, behavioural and neurological assessments, no practitioner can be very certain as to what they are dealing with or how best to treat it.

In the veterinary field, there is a long way to go to understanding and diagnosing neurological problems such as rage syndrome. It is a fascinating area which deserves *much* more attention.

Acknowledgements

I am extremely grateful to Maureen Adams, Laura Marsh, Gail Podberscek and Adrienne Thomas for their help.

11
Rehabilitation of Rescue Dogs and Cats

Julie M. Bedford

Two common misconceptions about animals available for rehoming at the many charity animal centres are that they have been rescued from a life of neglect and abuse or have behaviour problems that make them too dangerous or disruptive for the average person to manage. Thankfully neither is usually true. Most animals admitted to charities for rehoming are well cared for and much loved pets, which when obtained were intended to be with the family for life. In fact the most common reason for animals being handed over to one rescue organisation, The Blue Cross, is changes in the owners' circumstances. These include changes in work pattern, financial difficulties, divorce, house moves and illness/hospitalisation. The second most common reason for animals being brought to the Blue Cross is that they are strays. Almost 10 per cent of dogs and 30 per cent of cats handed into the Blue Cross are strays.

The least common reason for animals being handed to the Blue Cross is behaviour problems. This group accounts for less than 10 per cent of cats and 9 per cent of dogs, (Animals handed into the Blue Cross, 2002). Of those, most will have their behaviour problems resolved or managed. The majority are overcome by providing a more suitable activity, a change of environment or by placing the animal with an owner who due to his or her experience or expectations does not see the behaviour as a problem. So the majority of pets available at rehoming centres will be problem free and, with correct placing and introduction to the new home, will settle successfully.

Of the remaining few which do not fit into any of those categories most of them can be helped with sensible handling and a sensitive well informed approach to their management and rehabilitation.

What follows provides an insight into what is involved in the rehabilitation process, which is a guide for those working or aspiring to work within a rescue organisation. For the pet owner or potential pet owner it is an insight into the care taken where best practice is employed.

The rehabilitation process, for those which need it, has to begin when the current owner first approaches the animal charity to hand over the pet. Prior to, and at the hand over as much information as possible should be gathered. Comprehensive information helps to define the problem and the appropriate way to resolve it.

Taking a Full History

When the animal is brought to the centre for rehoming it is important that a detailed account of the animal's routine, management and behaviour are taken. This must be done without judgement – generally people do what they think will be the best to resolve their problem and taking a judgemental stance will prevent the client being honest and open. The client must be reassured and forewarned of the need to complete a questionnaire about their pet. The person collecting information should be genuinely interested in getting as much detail as possible but must do it sensitively and professionally. This process must be followed whether or not the animal has behaviour problems. The more information that is taken on an animal the more likely it is to be successfully rehomed.

What follows is a list of questions that should be asked – the list is taken from the admittance form used by Blue Cross Adoption Centres. The questions are only a starting point and should be developed and explored as the initial answers are given. Questions are better kept open (avoid yes and no answers) to ensure descriptive and qualitative information.

General Questions that Are Asked about Dogs and Cats

AGE?

SEX? IS IT NEUTERED?

BREED?

TEMPERAMENT DETAILS

These give a good indication of the overall behaviour of the animal.
How does he behave with strangers (men and women)?
How does he behave with children (0–4 years, 5–12 years, teenagers)?
How does he behave with dogs, cats?
How active is he?
How strong-willed is he?

How affectionate/tactile?
How does he behave when restrained/groomed?
Is he frightened of anything (noise, vets, cars etc)?
How does he behave at the vets?

GENERAL MANAGEMENT DETAILS
How long can he be left without problems?
Does he travel well?
Is he house-trained?
When and what is he fed. (Try to keep the dog on the same food and gradually change over to the brand used at the centre)

HAS THE ANIMAL BEEN SOCIALISED
How old was he when you obtained him?
Where did he come from (breeder – kept outside, domestic setting, etc.)?
Who lives in the home (adults/children/pets)?
How many homes has he had (including yours) – if he has had previous homes why did the previous owners give him up?
Is your home quiet, busy?
As a puppy/kitten did he meet many adults, children, dogs, cats? Did he have any unpleasant experiences with any of them?

THE OWNER/ANIMAL RELATIONSHIP AND WHAT TYPE OF INTERACTION THERE HAS BEEN
Have you ever needed to punish your pet – if so, how did you do it and what were the circumstances?
What makes your dog/cat happiest? – list three things
How responsive is he to commands?
Why are you giving your pet up for rehoming?

DOG SPECIFIC QUESTIONS

Did the dog have a lot of opportunity to play with other dogs when growing up?

How did he play with the other dogs? e.g., Which dog initiated play? Describe how they played, etc.

Where does the dog sleep at night?

Is he allowed on furniture?

Can you take bones, toys, and food away easily?

What is your dog's favourite game – tug of war, chasing, squeaky toys, won't play with toys, wrestling games with people?

How many times is your dog walked? – Is he off lead? How long are the walks

Has your dog ever growled, barked, snapped or bitten strangers (including visitors to the house) or members of the family and what were the circumstances?

Does your dog follow you from room to room?

How does your dog try to get your attention?

What does your dog do when left alone?

CAT SPECIFIC QUESTIONS

Is the cat

Confident?

Sociable?

Shy?

Does he prefer to be alone?

Can you give medication to the cat easily?

Does the cat like to be picked up?

Does he enjoy being cuddled, stroked?

How does your cat behave with cats outside the home?

How many other cats are in the home? How does he interact with them?

Does he use a litter tray (if so what litter is used, etc.)?

Is this shared with other cats?
Does he have a scratching post? Does he use it?
Is he used to going outside?
Can he use a cat flap?
Has the cat been used to living near a busy road?

The answers to these and follow-on questions are used to build up an individual profile. In addition to providing a guide as to the most suitable type of new home for the animal it can be used to identify problem behaviour, how and why it developed and how the owner dealt with it. A plan can then be developed as to how to manage the animal better and a programme devised to avoid or resolve the problem.

The information given by the owner is very important as it indicates how the animal behaves in a domestic environment, what it has learned and how it has been managed. But it must be supplemented by observations by staff whilst the animal is at the centre. Ongoing assessment and keeping a record of the initial and ongoing information where all staff dealing with the animal have access to it is essential for successful rehabilitation and rehoming.

Ongoing Assessment

This is important for all animals but particularly so for those with no known history, e.g. strays.

The ability to make predictions from assessment in environments such as kennels and catteries is limited and particularly difficult in cats (whose behaviour tends to be suppressed in new surroundings). Ongoing assessment can help to support the profile and identify problems that may require some work. To be valid the assessments cannot be done as an intensive set of kennel or cattery tests and observations (of the type where people introduce novel items or mildly intimidate to see what response they provoke). Although helpful, formal testing of this kind is artificial, in respect of context, and because they are a snapshot of the animal at a particular moment, tiring and stressful, and does not yield reliable results. Assessments are better and more realistic if carried out by appropriately trained staff during the normal routine of handling and caring for the dog or cat. For example, noting how the dog behaves when out for a walk or how well the cat tolerates grooming builds up a more accurate picture of the animal's responses and ability to cope with new situations. These 'real life' situations and are far more useful and predictive than contrived 'formal' tests.

Minimising the Effect of the Kennel or Cattery Environment

The contrasts between a home environment and a kennel or cattery are immense and seeing how well most animals cope with the transition is a testimony to just how adaptable and robust most animals are. However, some animals' behaviour will deteriorate whilst in kennels and catteries. This can be minimised by careful attention to their routine, environment and handling. A routine gives the animal predictability of 'what will happen when' and adds to the feeling of security. The kennel or cattery design should enhance this security and provide an area for the animal to retreat to if it is feeling overwhelmed or threatened. Physical comfort should also be ensured, temperature should be appropriate for the animal, the housing should be dry and draught free and noise kept to a minimum. Ensuring the dogs have a routine so that barking is restricted to times of excitement, e.g. feeding times, can reduce noise in kennels. Radios should not be too loud or left on all day.

Environmental enrichment should provide for the animals' behavioural needs. For example, cats need scratching posts, boxes to hide in, toys and high shelving. Dogs need toys; chew items and many enjoy extras such as paddling pools. Dogs also benefit from regular walks in a variety of different areas – pavements near traffic, fields near livestock. These activities not only contribute to the assessment of each dog but also prevent them from losing their familiarity with a variety of environments. Handling should be done daily, should be non-intrusive and gentle. Gradually these sessions should be built up to full grooming sessions and 'vet' like inspections. With basic routines and handling methods in place and consideration given to the animals' environment the next step is to work out what rehabilitation work needs to be done.

Making a Plan and Ensuring Everyone is Working Towards It

When an animal is identified as needing rehabilitation it is important to discuss and agree a plan of action for that dog or cat. Staff meetings when everyone can feed back their experience with the animal are a useful starting point. A plan of action must be agreed and then adhered to by all staff. Reminders such as notes on the animal's accommodation must be clearly posted. Progress reports must be kept to ensure that each individual takes responsibility in carrying out the programme. Progress should be regularly reviewed to ensure the rehabilitation programme is improving the animal's behaviour.

As members of the public usually have access to the animal areas they may need to be advised as to how to behave around a particular animal. If constant supervision is not possible, then instructions must be clearly displayed. For instance, an animal that is intimidated by people staring at it may have a sign displayed asking people not to stare – and most importantly why they should not stare. An alternative behaviour must also be given such as looking past the animal and speaking softly. If people understand why they have been asked to do something, and given an alternative behaviour, they are much more likely to comply and not hinder the progress of the dog or cat.

The public can even form part of the rehabilitation plan. Where animals are timid but not aggressive, a small holder can be attached to the front of the kennel or cattery and filled with food treats. A notice can be attached asking people to throw a tit bit in as they pass. This helps the animal associate strangers with a treat rather than intrusion. With young or boisterous dogs this can be taken further. The food treat should only be given if he is sitting quietly by the door (this should be reinforced by staff before relying entirely on the public). This is surprisingly effective and engages the public as well as the animal.

Having fishing-rod-type toys available in the cattery encourages the public to spend time playing with the animals in a non-intrusive way and can be done through the mesh of most cattery fronts. In addition to encouraging the animal to form pleasant associations with strangers these methods also assist in keeping the animals occupied.

Occupying the Animals, and Training in the centre
Bored and under stimulated animals are likely to invent ways of occupying themselves. Usually the things they find to do are not what pet owners want. Cats and dogs both benefit from being taught how to play with toys. Not only does it channel their energy and provide an outlet for some of their natural behaviours, it also assists in building bonds between the animal and people. Keeping the animals occupied and breaking up their day gives them something to look forward to and helps prevent their behaviour deteriorating. Feeding dogs at least twice a day helps them utilise their food more effectively but again adds an extra, positive encounter with their handler and an opportunity to train, e.g., asking the dog to sit before feeding. Feeding dry food via a food ball is something that cats seem to particularly benefit from. They are kept active 'chasing' their food, and this outlet for their normal hunting behaviour also appears to reduce some cats' aggression towards their handlers.

Short, regular, handling sessions for both cats and dogs are particularly useful – these should be pleasant and done at times other than those used to give medication, etc.

Basic training can also make a big difference – particularly for dogs. Teaching good manners, self-control and how to work for rewards not only makes a dog easier to rehome, but it channels their energy, develops alternative behaviours that can be rewarded, and prevents reinforcing unwanted behaviour. Basic exercises include waiting by doors, not jumping up, walking on a lead, sit, down and coming when called. Training must be consistent and implemented from the time the dog first arrives at the centre. Rewarding the dog whenever it displays the required behaviour will ensure that it will quickly learn what behaviours you want and which behaviours won't bring it praise, food or games. This need not take any additional time as it can be incorporated in the dog's daily walk. Walking dogs that are kennelled is essential – not only does the dog get attention on a one-to-one basis and a break from the kennel but, as mentioned earlier, it is an ideal opportunity for the handler to assess the dog and note how it behaves when it comes across traffic, strangers and other animals.

Minor Behaviour Problems

It is always important that any medical conditions are ruled out but problems not involving aggression can often be dealt with very effectively in the kennel or cattery. For example, attention-seeking problems are often resolved in rehoming centres as people are just not available to respond to behaviours exhibited to get attention. Providing this type of problem has been identified and the new owners advised when to give attention and when to withhold it, the problem is quickly resolved.

Spraying by cats, often caused by stressors – such as too many cats in the home, dogs, lots of visitors, etc, can be resolved by correctly identifying the problem (from the questions taken when admitted) and rehoming to an environment in which the cat is more likely to cope, for example, a quiet home with no other pets.

House-training problems can be resolved by providing the cat with a suitable covered, private litter tray and ensuring that a similar litter is used in the new home, placed in a quiet area easily accessible to the cat. Dogs that have not been house trained or their house-training has broken down can be rehomed with advice on appropriate house-training methods and ensuring that the dog has been well exercised and toileted before going into the home for the first time. Similar advice will also prevent the breakdown of house-training, as many animals are excited or nervous at being taken to a new home – in the excitement they may urinate or defecate in the home, especially if they have had a long car journey. This marks out for them a new 'toilet area' and in a new home the only way they can identify where to toilet is by smell.

It is also important to stress to the new owner to start off as they intend to continue. Many owners feel sorry for their new pets and indulge them in behaviours that may not be acceptable later on in the relationship. For example, dogs may initially be allowed on the furniture, but when they are settled in or when the weather is wet the owner decides that they can no longer be on furniture. Having to learn new rules is confusing for the dog and may cause conflict. It is better for all, speeds up the settling in process and prevents problems developing, if the animal has consistency from day one.

Nervousness

Nervousness is a common problem in dogs and cats that are in rehoming centres and it often stems from inadequate early life experience. Whilst that can never be fully overcome, a gentle desensitisation programme will start to help the animal readjust and cope with life. It is important that nervous animals are kept safe and relatively quiet – so place cats well away from the kennels and keep nervous dogs in quiet areas of the kennels. Avoid over-stressing the animal by having multiple handlers or intrusive practices. Initially just one or two members of staff should care for the animal.

Allow the animal to build a bond with them before introducing new people. Basic considerations are:

- Handling the animal at times other than when needing to do intrusive procedures such as worming and vaccination.
- Giving the animal a sense of control in its ability to move away and feel safe. So cats need a hiding place and high-up shelving. Dogs need an area they will not be stared at (simply achieved by throwing a blanket over the front of the kennel door so it is partly obscured and the dog can hide behind it). Animals which know that they have a safe retreat are much more likely to come forward and investigate the visitor.
- Trying to link the people and things the animal is afraid of with something the animal enjoys – so a stranger may engage the animal in play or offer a particularly tasty food treat, as discussed above.
- Allowing the animal to approach rather than approaching the animal – this takes longer initially but by going at the animal's pace it will help the animal to develop its confidence and learn to trust people. Progress is then much quicker.
- Teaching the animal alternative behaviours, e.g., giving a paw, lying down, etc., is also useful – nervous and fearful animals really appreciate being given a different way of coping – but only if you ensure they are not threatened or overwhelmed. If the animal does not trust you it will have to rely on old behaviour patterns to cope with new situations.
- Training a new behaviour is something that is often considered and done with dogs but rarely with cats. Cats in particular respond very well to clicker training

and now that clicker training is becoming more popular and more people are becoming competent at using it, it is an ideal method for retraining many cats.

Aggression

Rehabilitating aggressive animals depends on the type of aggression, how long the animal has been aggressive and how severe the aggression is. If staff, public or other animals are seriously at risk then the decision may be taken not to attempt rehabilitation. However, many types of aggression may be avoided and controlled by careful rehoming and advice.

Some aggression, particularly in older cats, may be attributed to pain. After veterinary examination, careful handling and treatment for painful conditions such as arthritis often transforms the cat's behaviour.

Cats aggressive towards other cats in the home can be rehomed singly and in an area where there are few other cats. (Despotic cats are an exception as they may travel considerable distances to find and attack other cats). By not being forced into close contact with other cats the cat will be able to avoid encounters with them and general aggression levels will fall. Cats may also be aggressive towards their owners – this can be for a variety of reasons but most can be reduced by ensuring that the cat has sufficient stimulation – access to outdoors or appropriate play with a fishing-rod-type toy (these are particularly good as the owner can keep a safe distance from the cat). Owners can also be taught to recognise early warning signs and management techniques. Cats which bite or scratch when picked up can be taught to jump on the owner's lap by luring them up with a prawn or tuna fish. Watching out for body language and facial expressions that accompany aggression, such as twitching skin, tail flicks and body stiffening will enable the owner to move before the cat tries to move them.

Dogs which are aggressive towards other dogs can become much worse in kennels, but if there is the option of isolating them (for example, keeping them in the staff room) and teaching them to bond strongly with humans, then a great deal can be done at the centre. When a dog has developed a strong bond with people it can be gradually exposed to other dogs in a controlled and safe way. Additional control can be achieved by using a head collar that will give the handler the opportunity to divert the dog's attention and prevent him from lunging towards other dogs.

Dogs which are accustomed to controlling their owners and demanding what they want, and have become uninhibited can be sent out with modification programmes advising how to manage the dog and ensure that it works for rewards and is not placed in a situation where it can guard resources.

In all cases follow-up advice on behaviour and support can help the new owner to resolve the problem (the new owner must be fully informed of the problem prior to adoption). It is the ongoing support that ensures the owner is motivated and confident enough to resolve the problem.

Careful Selection and Matching of New Owners

Matching animals with new owners takes a lot of time and effort, but not doing it is a false economy. Getting the right match saves a great deal of time, as the same animals aren't returned for 'recycling'. It is better to refuse a prospective owner an animal which is not suitable and keep it for a while longer, than place it just because it is better in a home rather than a rescue centre. By fully explaining the decision to a perspective owner and identifying a more suitable animal for him or her, you have a better-educated owner and have increased the chance of the adoption being successful.

Of course it is important that we do everything to rehome correctly the first time around. Failing to do so costs time and money which charities can ill afford and more importantly has a negative impact on animal welfare. Moving from home to home and being brought back into kennels/catteries – however well run – are unsettling and distressing for dogs and cats. In all cases follow-up advice on behaviour and support can help the new owner to resolve the problem.

Strays

The history of these animals is unknown and additional precautions will be needed when rehoming. For example, it is wise, in order to prevent problems developing, not to rehome adult animals to homes with young children. This is because the dog or cats experience with children is not known and it is unfair to the animal and family to risk placing a dog or cat into what could be a very difficult environment for that individual animal.

Ongoing Support and Advice

The unsettling experience of changing homes and the different expectations of different people can lead to behaviour problems developing which the new owner will find problematic. These cannot always be foreseen but if the client has been made to feel comfortable during the adoption process and information has been freely available, they are much more likely to seek and value advice given by the rehoming organisation. The new owners are also much more likely to seek help as soon as the problem becomes apparent, rather than waiting for it to become unmanageable. In the early stages most behaviour problems can be explained and resolved by giving good quality information via the telephone or information leaflets. Some problems, especially those involving aggression will need much more input, including regular visits and individually prescribed modification programmes. Taking this sort of action in good time can prevent new owners having to return the animal.

Taking an interest in each animal and person arriving at a rehoming centre, gaining and passing on as much relevant information as possible about them will ensure the best possible match is made. This combined with ongoing support in the form of good quality information, openness, help and advice ensures a high standard of animal welfare and customer care. Working towards these principles enabled The Blue Cross animal welfare charity to have less than 4 per cent of dogs and 1.5 per cent of cats rehomed by the charity, returned.

12
Behavioural Problems in Ageing Pets

Sarah Heath

Introduction

As the state of health care continues to improve, more people are having to deal with the varied implications of old age, and society is recognising that the ageing population has specific needs. In a similar way the pet population is steadily growing older and the proportion of pets which qualify as geriatrics is increasing. With this change in population comes a need to recognise the specific requirements of this special group of animals, in terms of their behavioural and emotional needs, and to offer appropriate support for their owners.

Age as an Inevitable Process

Ageing is one of those inescapable facts of life and veterinary surgeons are no more able than their medical counterparts to offer any miracle solution to the issue of old age. However, there is now a considerable body of information available about the changes that occur during the ageing process and about how these changes can be affected by diet, lifestyle and medication. With this knowledge at our fingertips we can certainly increase the quality of life for the ageing pet population and ensure that pets and their owners get the most from their relationship at this time.

A Changing Population

Advances in veterinary medicine over recent years have meant that old pets are now offered the very best in medical care during their later years and many will enjoy excellent physical health into their late teens and even beyond. However it is also important to remember that ageing affects not only the major organs such as the heart, lungs, liver and kidneys but also the nervous system and more specifically the brain. Changes in neurotransmitter levels, alterations in membrane permeability and increased production of free radicals within the central nervous system can all lead to the onset of age-related behavioural changes, and when pets start to show symptoms related to these changes they begin to require a very specific form of veterinary attention. Of course behavioural changes in elderly patients are not always related to specific issues of brain ageing, and the potential involvement of medical conditions such as pain, failing sensory systems, compromised blood flow to the central nervous system and age-related deterioration in functioning of the major organs, such as liver, kidneys and heart, all need to be considered. In addition, environmental and psychological factors should not be overlooked, and it is important to remember that older animals may be less adaptable to change in their social or physical surroundings and may show marked behavioural changes in response to changes in their household or routine as a result.

Slowing Down or True Brain Ageing?

As the ageing process takes its toll on a dog's heart and brain some changes in behaviour and personality are almost inevitable and it is important to be able to differentiate between cases where the animal is simply slowing down and those where the animal is finding it increasingly hard to function at a social level. In some cases behavioural changes in old dogs will noticeably resemble the symptoms of senility or Alzheimer's disease in people. Lack of connection between behaviour and context is a classic sign of brain ageing and many owners have reported that their pet seems like a stranger in its own home. In cases of cognitive dysfunction in dogs and cats post-mortems have shown similar neuropathological lesions to those seen in people with Alzheimer's disease and there is no doubt that this condition is part of mainstream veterinary medicine.

Recognising the Process of Brain Ageing

Cognitive dysfunction (CD) is the name given to the age-related progressive neurodegenerative disorder which is characterised by specific behavioural changes, and physical modifications of the central nervous system, such as the deposition of beta amyloid plaques, production of free radicals and onset of vascular changes, which lead to characteristic neuropathololgy.

Unlike its physical counterparts cognitive dysfunction is characterised by a lack of recognisable clinical symptoms and this can lead to difficulty in identifying cases, especially in the early stages of the condition. Patients suffering from CD primarily exhibit changes in their behaviour, and unless owners are asked the right sorts of questions it is easy for the disease to go undetected. Alterations in learning and memory, together with changes in social interaction, sleep/wake cycles and activity patterns are traditionally associated with the condition, and specific changes in the perception of and response to stimuli in the environment, in levels of anxiety and restlessness and in spatial orientation and levels of confusion are also identified.

In order to assist in the clinical evaluation of these cases the behavioural changes associated with this condition are commonly divided into four main groups, namely:

- disorientation
- changes in social and environmental interaction
- changes in sleep/wake cycle
- changes in response to learned associations and commands.

On their own each of these categories could be indicative of a physical disease or could be the result of a purely behavioural condition, but when signs are present from each of these categories, and most importantly from the first two, a diagnosis of cognitive dysfunction needs to be considered.

The Human Dimension

A study by Neilson and colleagues in 2001 showed that 28 per cent of dogs in a population aged between eleven and twelve years and 68 per cent of a population aged between fifteen and sixteen years showed at least one sign which was consistent with a diagnosis of canine cognitive dysfunction and yet only 12 per cent of pet owners reported these changes to their veterinary practice. One of the reasons for this discrepancy may be that many owners are reluctant to speak to their veterinary practice for fear of being advised to consider euthanasia for their beloved companion. As a result they tolerate changes in their pet's behaviour, seeing them as an inevitable effect of growing old, rather than a symptom of a underlying disease process that can be managed, even if it cannot be cured, and the belief that nothing can be done to change them will drive many owners to make considerable changes in their own lifestyle in order to accommodate their pet's peculiar ways.

Alterations in sleep/wake cycles can take the largest toll on the humans in the household, but owners will often put up with broken nights and significant sleep deprivation rather than approach their veterinary practice and run the risk of being persuaded to put their pet to sleep. Breakdown in house-training can also put an enormous strain on the pet-owner relationship but many owners feel guilty about their negative feelings toward their pet and chastise themselves for not being patient with their ageing companion at these difficult times. After all, quality of life is a difficult concept, and it is important to remember that the criteria by which owners judge the quality of their pet's life will be very different from those that a detached professional might employ. The pet is a valued member of the owner's family, and its behaviour is often one of the major indicators of the strength of that relationship as well as one of the most significant sources of strain upon it. However, in the context of old age, forgiveness for changes in sleep patterns and breakdowns in house-training is usually forthcoming from owners who see these as an inconvenience that simply has to be endured.

In contrast changes in social interaction and signs of disorientation are not seen as inconvenient alterations that need to be accommodated, but rather as indicators of the pet's distress. Owners find it upsetting to see their companion looking confused and when their dog fails to respond to previously known commands, or their cat startles when they enter the room as if he does not recognise them, the effect on the pet-owner relationship can be devastating. Feelings of guilt at the prospect of prolonging life for a pet that is no longer fully aware of its surroundings can be very real but belief that euthanasia is an unacceptable way out can lead to emotional conflict that owners find hard to resolve. Some find it hard to forgive themselves for even considering euthanasia when their pet appears to be in such good physical health, since human society frowns upon the concept of euthanasia and the only way of justifying it in a veterinary context is as a means of stopping physical suffering. Putting to sleep due to changes in behaviour is therefore seen by many owners as a convenience and this only adds to the guilt.

Ageing Felines

Although most of the companion animal literature relating to behavioural changes in old age have been concerned with the dog, some work on the effects of ageing on feline behaviour has also been carried out. In a prevalence study conducted by Landsberg and colleagues in Canada, 35 per cent of cats over eleven years of age demonstrated signs of cognitive dysfunction which were not complicated by any other medical condition. In the category of cats older than fifteen years the percentage of affected cats increased to 48 per cent and these cats also showed more signs of the condition than their younger counterparts.

The presentation of cognitive dysfunction in a feline context is very similar to the condition in dogs and to related conditions in people, with signs of confusion, disorientation, alteration in social behaviour and breakdowns in house-training all being reported. Night-time waking, excessive vocalisation and aimless wandering and pacing are also commonly cited in the ageing feline population and changes in sleep/wake cycles appear to be the most commonly reported change in the population of cats older than fifteen years.

Asking the Right Questions

Noticing the signs of cognitive dysfunction early on is crucial and there is a much better prognosis in terms of the level of improvement and the extension of good quality life if treatment is instituted in the early stages of this disease. However, it is this early stage that often goes unnoticed and the most effective way of increasing the detection rate for this condition is for elderly pets to have regular veterinary examinations, which include a behavioural as well as a physical health check. Information about the animal's social behaviour, as well its sleep patterns and toileting habits, will lead to early detection of the condition and allow treatment to be begin as quickly as possible.

Investigation of Canine Cognitive Dysfunction

In order to assist in the early detection of cognitive dysfunction in dogs it can be useful to divide the areas of investigation into the four main categories and then to ask questions which are designed to identify alterations in behaviour. A similar approach will be needed in the investigation of feline cognitive dysfunction, but at present far more is known about the subtle signs seen in dogs and information about treatment opportunities is certainly more readily available in the canine field.

DISORIENTATION

Dogs suffering from cognitive dysfunction will show a delay in the recognition of people, places and objects and in some cases there may be no recognition at all. Obviously when dogs fail to recognise their owners this is likely to be noticed quite quickly, but failure to respond to people who call regularly at the house or to those that are met while out on daily exercise may go unnoticed for quite some time. The way in which the dog greets people at home and on walks can help to identify canine cognitive dysfunction in its early stages. Some old dogs suffering from cognitive dysfunction will wander aimlessly around the house and it may be obvious that these individuals are disorientated and confused, but in other cases more subtle alterations in the dog's reaction to previously familiar objects, such as household furniture and trees and bushes in the garden, can help to pinpoint the symptoms of cognitive dysfunction. These dogs will often bark at these objects as if they have never been seen before and some may show real fear. Failure to recognise the home driveway on the return from a walk or a tendency to sit at internal doors when asking to go out into the garden can also be symptoms of disorientation. Unexplained staring is another potential indicator of this condition.

CHANGES IN SOCIAL AND ENVIRONMENTAL INTERACTION

One of the most obvious signs of a lack of connection between context and behaviour is seen in the interactions that older dogs have with the people and other dogs that they come into contact with. One of the most distressing examples of this is an alteration in the social interaction between the dog and its owners and a decrease in the enthusiasm of greeting behaviour. Together with a decrease in the time spent engaging in play and in affectionate interaction this can signal the onset of age-related brain changes. There is often a change in the consistency and speed with which these dogs respond to commands, and owners often mistake this change for stubborn behaviour. When considering interactions with other dogs, it is not uncommon to see an increase in confrontational reactions in other dogs which appear to be threatened by the bizarre behaviour of the geriatric individual. Dogs suffering from cognitive dysfunction may also become more irritable themselves and owners may be aware of an increase in aggression from their dog together with a general decrease in the desire to interact and play with other dogs on walks.

CHANGES IN SLEEP/WAKE CYCLES

Alterations in sleep/wake cycles are common in cases of canine cognitive dysfunction but, unless the dog is disrupting the owner's sleep, this aspect may easily go unnoticed. When the dog is whining or barking at night most owners will rapidly respond by taking their pet out into the garden on the assumption that it needs to relieve itself but dogs suffering from cognitive dysfunction will rarely need to eliminate on these occasions. In some cases they may respond to being outside by toileting but once back in the house they will be very slow to settle and will soon recommence barking and whining. Pacing is another common feature in these cases and owners often comment that their dog shows signs of agitation and restlessness at their usual bedtime and paces and vocalises when they make preparations to go to bed.

Waking in the middle of the night is another classic symptom of cognitive dysfunction and since these dogs are also disorientated and confused they will often seek out their owners when they wake. If the dog is housed downstairs during the night this may cause the dog to scratch at doors and this symptom can commonly lead to confusion with separation anxiety. During the day the sleep/wake cycle is also affected and dogs suffering from the behavioural effects of brain-ageing will sleep for longer in the day but once again this symptom may easily be overlooked and considered to be a normal change in a dog of advancing years.

LOSS OF PREVIOUSLY LEARNED BEHAVIOURAL RESPONSES

Breakdown in house-training is probably the most documented example of this change in behaviour and can occur in cases of canine cognitive dysfunction for two reasons. First, the disorientation associated with the condition can lead to a situation where dogs sit at internal doors when they want to go out into the garden and owners can easily disregard this signal until it is too late. Staring at the hinged side of the door rather than the handle side has also been noted by owners and in these cases they often observe the peculiar behaviour but do not recognise it as a signal for needing access to outdoors. Another factor in the house-training problems associated with old age behavioural changes is a failure to maintain associations with suitable latrine substrates, and in many cases there is a history of a progressive breakdown in associations leading to elimination in a wide range of unsuitable locations. These dogs often start by not always toileting on grass as they have done in the past and making mistakes on the patio, on the flowerbeds and eventually on the carpeted floors in the house. In many cases the onset of this problem behaviour is gradual and it is only when the dog is toileting in a number of locations that the owners realise that these are not one-off accidents.

MEDICAL DIFFERENTIALS

In all the categories of behavioural symptoms for cognitive dysfunction there are a number of medical conditions that need to be considered as potential causes. Deterioration in sensory functions, such as sight and hearing, need to be considered when signs of disorientation, changes in social interaction and changes in sleep/wake cycles are encountered, since an animal which is not fully aware of its surroundings through a lack of sensory input can easily present with signs which resemble those seen in cognitive dysfunction. Cardiovascular and neurological disorders will also need to be ruled out and the influence of pain on activity levels, social interaction and sleep patterns should be investigated. Disorders of the gastrointestinal and urinary tracts will need to be considered in cases where a breakdown in house-training is the major presenting sign and medical conditions which result in increased thirst or increased urination will also need to be included in the investigation. Only once a thorough medical investigation has been carried out by a veterinary surgeon can the behavioural investigation continue and treatment for cognitive dysfunction be considered.

Treatment for Age-related Behavioural Changes

Most of the proactive research into the treatment of cognitive dysfunction has been carried out in dogs and at the moment medications and specifically formulated diets are only marketed for dogs suffering from this condition. However, recent work in Canada has suggested that a similar therapeutic approach, consisting of a combination of dietary management, medical treatment and behavioural therapy, is just as applicable to the feline population.

Medication for Cognitive Dysfunction

Although primarily behavioural in presentation, there can be no doubt that cognitive dysfunction is a medical condition, and when deciding on medication to treat these cases it is important to consider the changes that are occurring within the body systems, including the central nervous system. During the normal ageing process behavioural changes can result from changes in the blood flow to the brain, and medication with drugs which enhance the availability of good cerebral circulation can vastly improve this situation and give increased energy and vigour to dogs which are beginning to slow down. In addition, there are drugs which lead to a protective effect on the central nervous system and these may be beneficial in cases of canine cognitive dysfunction where internal changes such as a depletion of brain dopamine levels and an increase in the presence of free radicals, leading to cell injury and brain pathology, need to be considered.

In the United Kingdom and the USA there has been a lot of publicity relating to the use of selegiline hydrochloride as a therapeutic approach to cognitive dysfunction and three important actions have been identified in relation to the treatment of these cases. First, selegiline enhances brain dopamine concentrations and metabolism, secondly, it decreases substances in the brain which are responsible for neural cell damage; and, thirdly, it protects nerve cells, decreases cell death, and promotes synthesis of nerve-growth factors. It is administered as a once-daily dose and in cases of dementia the dog will usually require long-term medication. Then onset of action of the drug is not immediate and it has been reported that clinical changes have been delayed for up to six weeks after commencing treatment. However, others report changes within three weeks. In long-standing cases of canine cognitive decline it has been reported that the effects of single drug therapy may not be sustained and that such cases may respond better to combination therapy, but in such situations it is important to discuss the dog's behaviour in detail with a veterinary surgeon who can monitor responses to treatment and modify treatment as appropriate.

A Nutritional Approach to Cognitive Dysfunction

It has long been accepted that nutrition can play a vital role in the treatment of organic disease and in cases where behavioural problems in the geriatric pet are attributable to conditions such as renal failure, the use of prescription diets should obviously be considered. However, recently there has been increasing interest in the role of nutrition in the treatment of age-related progressive degeneration of the central nervous system and in the clinical consequences of that process. In particular research has looked at the effects of dietary manipulation on canine short-term memory and has assessed this with reference to particular learning and memory tasks. This is particularly significant in relation to treatment of canine cognitive dysfunction, since loss of short-term memory is known to be one of the first indications of similar conditions in humans. Researchers have found that cognitive performance can be improved with a diet supplemented with a broad spectrum of anti-oxidants, which are believed to prevent the development of the age-related neuropathology. In addition anti-oxidants are believed to promote recovery in neurons that are exhibiting signs of neuropathology and therefore commercially available diets and nutritional supplements, which are enhanced with these agents, are believed to offer another option in the treatment of behavioural disorders in the older pet. The role of nutrition in the prevention and treatment of both canine and feline cognitive dysfunction is currently the subject of extensive research and it is likely that further developments will be seen in the near future.

Providing Behavioural Support for a Medical Condition
Although cognitive dysfunction is undoubtedly a medical condition there is a strong behavioural component and behavioural therapy is needed as a support for the pharmacological treatment. One of the consequences of age-related behavioural disorders is the loss of learned responses. As a result dogs may lose their ability to perform simple tasks or respond to previously known commands and cats may show a breakdown in house-training or a reluctance to respond to being called in from the garden. Obviously re-establishing these learned responses is particularly relevant in treating cases of canine cognitive dysfunction. Teaching these dogs old tricks, however, will involve a large amount of patience and understanding. The use of simple

unambiguous commands and clear reward signals is essential. Ideally rewards should be things that the pet particularly values and this will be dependent on the breed and will also vary from individual to individual. For example, some dogs will value games, others petting and others food. The use of a clicker, which has been previously associated with reward in a simple introduction process, gives the pet a clear unambiguous signal that will help to reinforce success. Patients will often need to be taught commands from scratch, and house-training and introduction of very basic obedience commands will be part of the process. The aim of behavioural therapy in these cases is to give structure and predictability to the environment, in a way which helps the dog to understand what is expected. Cognitive dysfunction results in a lack of association between action and context, so it is important to ensure that all commands are consistent and that the dog is given as many clear signals of success as possible. Visual signals can be very useful, provided the dog is not showing any sensory deficits of course, and marking exit doors can be beneficial for those dogs whose disorientation is leading to house-soiling problems. Cognitive-dysfunction patients often show difficulty in concentrating and therefore it is important to introduce games that will provide mental stimulation and increase interaction between pet and owner. Ideally play and exercise sessions should be kept short and involve simple tasks, which are repeated frequently and culminate in a positive reward for the pet. For example, several short exercise outings each day will be preferable to one long one since this will stimulate the pet's interest in the environment and provide increased opportunities for interaction between the pet and its owner.

The Effect of Treating Cognitive Dysfunction on the Pet-Owner Relationship

The onset of old age is an inevitable fact of life and for many pets their transition into the ranks of the geriatrics is smooth. However, this is not always the case, and when signs of disorientation and confusion begin, many owners find it hard to recognise their faithful family friend. The changes in social interaction can make brain-ageing a distressing condition for the owner and when dealing with these cases it is important to remember the human element. Detecting the symptoms of this condition at the earliest opportunity will enable pets to receive appropriate veterinary care and maximise the benefits of therapy in terms of increased quality and duration of life.

13
Issues in Companion Animal Welfare

Donna Brander and Natalie Waran

'Issues of companion animal welfare are often too close to home to be seen by members of the general public.'
Professor Bernard Rollin – Animal Ethicist

Attitudes Towards Animals

The human animal bond is considered to be over 12,000 years old and was probably originally based on a mutually beneficial relationship. Dogs and cats exploited the tribal life of human beings as scavengers for the dross of the nomadic wanderers. The dogs would defend this resource from outsiders and thus became a helpful alarm against the approach of men or beasts. Cats were a natural vermin control. Sometimes dogs would be used as a beast of burden for nomadic peoples, and both dogs and cats could ultimately be used as food. But there is evidence that, even from the beginning, these animals were sometimes kept solely as companion animals.

Many tribal people (for example, Native Americans and Aboriginal people) believe animals are individuals. They believe that animals are thinking, feeling beings that are worthy of respect as ancestors and relatives. Some tribes of Native Americans call them the 'animal people' and consider animals to be part of the circle of life. Interestingly, many of the game laws, closed seasons and limited harvesting of animals that we apply today have roots in tribal thinking.

Various other ways of considering animals have developed through the centuries. Dominionism dictates that animals exist to serve humankind. Stewardship perceives humans as caretakers of other animal life. Evolutionary Perspectivism is problematic for some, as it looks only at the development of life from a human point of view. Critical anthropomorphism recognises the problem of homocentricism and attempts to assess behaviour in animals critically.

Religious beliefs have also played a crucial part in attitudes towards animals. Some religions could almost be considered to be animal liberationist. Buddhism, Jainism and the yogic branches of Hinduism perceive animals as having rights, and have a basic tenet of kindness towards all animals. The Quakers were the first Christian sect to oppose blood sports and incorporate kindness to animals as an article of faith.

Attitudes towards animals have changed throughout history. Early Christian attitudes considered man to have dominion over all animals. In the Middle Ages, animals were used for work, food and skins, but by the late Middle Ages the line between humans and animals became more blurred, to the extent that animals were often tried for crimes just as if they were human beings.

In the eighteenth century, reason and intellect set humans apart from animals. The Cartesian view of animals as machines led indirectly to the belief that animals could feel no pain and, therefore, there was no such thing as cruelty to animals. Cruelty to animals infused eighteenth-century Britain, and the ill-treatment of animals was a part of social fabric. However, by the nineteenth century, humans began to define themselves as creatures of feeling and passion, rather than intellect, and they therefore became more empathetic towards animals. It becomes clear that what we think about animals largely depends on how we define ourselves.

In the nineteenth century attitudes towards animals began to change dramatically with the theories of Darwin. Once again, there was a joining of humans to animals. With these philosophies came what is known as the Generosity Paradigm, which suggests that the vulnerable, unprotected, undefended and morally innocent have a moral priority over other competing claims. Along with this thinking came the concept of protection against cruelty towards animals, and organisations such as the American Society for the Prevention of Cruelty to Animals (1866) and the Royal Society for the Prevention of Cruelty to Animals (1824) were formed.

What Is Welfare?

There are various definitions of animal welfare but, in general, an animal's welfare relates to its ability to adjust to, or cope successfully with, the prevailing conditions in which it finds itself. As a concept, welfare can be described in terms of the Farm Animal Welfare Council's 'Five Freedoms'. These state that the basic requirements for any animal are food, water, good ventilation and protection from risk of injury or disease. In addition animals should be able to perform their natural behaviour and be free from suffering.

Most pet owners equate welfare with the quality of an animal's life. The problem is that quality of life can be difficult to determine accurately, since there is no one simple indicator that is universally accepted as a measure of welfare. There are a number of measures that can be used, ranging from measures of the animal's biological functioning, for example, its health and reproductive ability, to more subjective assessments of its emotional state, including how free it is to perform natural behaviour. In general most owners tend to presume that their dog is experiencing positive emotional states, such as happiness, when it behaves in a certain way towards them, for example, wagging the tail, and showing play behaviour. In this way owners often describe their dog or cat's welfare in anthropomorphic terms. Interestingly, people rarely see themselves as restricting the freedom of their companion animals, for example, by placing them in an unnatural setting or restricting their interactions with conspecifics. There is also no attempt at objective assessment of whether or not these animals are free from unpleasant emotional states.

Sadly, it is often the case that the care of pet animals can be misguided, although well intentioned. It is still all too common to consider the care and welfare of animals as being based on a 'common sense' approach. This approach can often lead to pets experiencing levels of care and management that fall short of their basic requirements.

Welfare problems arise when environmental conditions are so extreme that the animal can no longer cope successfully. In dogs and cats such as an inability to cope may manifest itself in destructive behaviour, aggression, increased sensitivity to noise, attention-seeking behaviour or stereotypical responses. These problem behaviours must be considered as indicators of a welfare problem. What is therefore required is a method of assessment of animal needs that is not homocentric.

Animal Welfare Legislation

In the UK, animals were traditionally viewed as property, and as such were protected because they belonged to the owner. In 1809 in the UK, Martin Erskine presented the first Bill to Parliament that was concerned with the protection of animals against cruel acts. Although this failed to become law, it did encourage others to consider the matter worthy of further debate. It was, however 1822 before Martin's Act finally became law, making it an offence for any person to beat, abuse or ill-treat any horse, or other beast of burden, and a further four years before dogs became protected under the Act. By 1840, the principle of animal welfare legislation had become accepted, and The Prevention of Cruelty to Animals Act became law in England in 1849. In 1911, the existing anti-cruelty legislation was consolidated and extended by the passing of the Protection of Animals Act. Since then there have been nine amendments to the Act. In essence the Act makes it an offence to cause any 'unnecessary suffering' to any domestic or captive animal.

The problem with this Act is that, as with most legislation, it is concerned with extremes. The statutes tend to focus on animal cruelty, and as such they do not seek to promote higher standards of welfare. Interestingly, in the case of companion animals, UK law is less advanced than the Council of Europe's Convention on pets. The European Convention for the Protection of Pet Animals is based on a 'duty of care' philosophy such as is described in the welfare codes for farmed animals. The proposed new UK Animal Welfare legislation based on the European convention has yet to make its way successfully through Parliament.

There are various other pieces of UK legislation that are

more specific and aim to control the use, housing and breeding conditions for companion animals. For example, the control of dog breeding is covered by The Breeding of Dogs Act 1973, 1991 and the Breeding and Sale of Dogs (Welfare) Act 1999. Under this legislation, anyone who is in the business of breeding and selling dogs requires a licence. This is granted only if the animals in the breeding establishment are suitably accommodated, fed, exercised and protected from disease and fire. It is for local authorities, which have extensive powers, to check on the standards of health, welfare and accommodation of the animals, to enforce the requirements of the Act. In addition the 1999 Act provides that bitches may not be mated until they are at least one year old and that they give birth to no more than six litters in a lifetime, and no more than one litter per year.

The Dangerous Dogs Act 1991 is another example of legislation that has welfare implications for pet dogs; however in this case the issues lie with the drafting of the legislation. After a series of well publicised attacks by dogs on humans, the UK Government hastily responded with an Act that, amongst other things, aimed to control the breeding and movements of certain breeds of dogs. The confusion lay in the interpretation of the wording of the Act, which led to long delays in proving whether a dog was of a certain type, and, as such, whether the owners were subject to prosecution for having custody of it, off-lead or un-muzzled, in a public place. The consequences of this were that many dogs were confiscated from their owners and placed in kennels, with the average length of detention being almost two and half years. At one point the Metropolitan Police were holding some 448 dogs under this Act.

Of course animal welfare law is constantly being revisited and revised and new proposed legislation is constantly being put before Parliament. For example a recent controversial Bill relates to the sale or use of collars for animals, which administer electric shocks.

Common Welfare Problems

Typically companion animals are required to fit their owner's lifestyle and additionally, the owner's concept of a social system. Human misunderstanding and high expectation is usually the reason for the development of problem behaviours. Of course companion animal behaviour problems usually impact on the animal's human companions and the behaviour is either addressed or the relationship abandoned. Dog and cat rescue homes provide temporary care and a rehoming service for thousands of pets each year, and one of the reasons for animals being placed in such homes has been shown to be behaviour problems.

PROBLEMS FOR DOGS

Dogs have a different social system to humans, and may have problems of separation from their human 'pack' unless accustomed to the situation. The 'urban dog' is an example of the social pack animal that is left alone without its group members. This causes the animal distress and it may respond by barking, howling or whining in an attempt to deal with the situation. The dog may attempt to escape the confinement or, in extreme distress, urinate, defecate or engage in destruction. Preventing such problem behaviour is the most effective way of avoiding the stress that this imposes upon the dog. During early development it is suggested that the dog be trained to deal with such unnatural situations in order to be comfortable with human lifestyle requirements.

Pet dogs can also exhibit problems with anxiety that can lead to aggression or fearful behaviour. This sort of problem behaviour can be attributed to the lack of human understanding of the different (linear) hierarchical social system required by dogs as compared with the human social system. It can also be caused by a lack of understanding of the importance of exposing puppies to commonly encountered stimuli. Problems arise where the dog is treated as an equal in a relationship, or is expected to share its food bowl, or some other important resource. A dog that is unwittingly allowed to develop a sense of high status may become competitive, which may lead to agonistic behaviour, whereas a low-status dog may exhibit signs of increasing fearfulness, such as hiding behaviour.

PROBLEMS FOR CATS

Problems of territory are often the cat's main welfare problem. It has been stated that the single most stressful situation for a cat is another cat. This can lead to problems when a household or neighbourhood has many cats. Even one cat, chosen to keep the resident cat 'company', may cause stress leading to behavioural problems, such as aggression between the cats, aggression towards people and marking in the house by one or both cats. Cats *are* social and many enjoy having others of their species around. But they prefer to choose their own companions and they may not necessarily accept one that has been arbitrarily brought into their life by well-meaning people. Most cats require many options to live within a territory in peace. A cat's life can be very complex in regard to time spent in various areas. Availability and location of resting areas, movement from one area to another, how food and water are dispensed and where they are dispensed all have to be considered from the perspective of their natural behaviour. For example cats like to snack rather than being fed on a schedule.

An interesting debate that often arises is whether cats require access to outdoors. In the UK, there appears to be a belief that cats need to be able to get outdoors, and, even in urban areas, cats may be shut out of the house during certain periods of the day. Often these time periods relate to the owner's lifestyle and not to the cat's, and this can lead to welfare problems where the cat is unable to avoid unwanted interactions or close encounters with neighbourhood cats. A cat's lifestyle can have a significant effect on its chances of being killed. A study carried out in Cambridge found that in a sample of 128 cats involved in road traffic accidents, younger male cats which spent more time out of doors during darkness were predisposed to such accidents. Neutered females, pedigree cats, and cats kept inside at night were less likely to be killed.

Cats often have specific litter-tray requirements such as placement of trays, number of trays, type of trays, and type of litter. Many of the needs of cats are simply not part of the human sensibility, and a 'common sense' approach to keeping cats will, at best, only be adequate, and may not address the animals' behavioural and environmental welfare. It is interesting to note that in the not so distant past, it was thought that cats did not feel pain to the same degree as other animals. This judgement was based on the cats' response (or lack of response) to stress and pain. As a prey animal, part of the cat's survival repertoire is to remain still and hidden when stressed or in pain.

Cats commonly hide when they are distressed, and some cats spend a lot of time hiding in the household rather than interacting with family life. Some of these animals may be in extreme distress, or even pain, but may be neglected because this behaviour is perceived as normal for the cat. While it may be normal for cats to be more discreet in their responses than a dog, any cat which spends a lot of time hiding or running away should be considered as behaviourally distressed, or in pain, rather than behaving normally. Lack of knowledge of a species causes inappropriate expectations. Many people either think cats are just like dogs, and treat them accordingly, or they have a misconception of true cat behaviour. Either way, the welfare of the cat is marginalised.

Feeding

It is interesting to note that the general public does not consider overfeeding animals, whether cats or dogs, a welfare issue. Most veterinarians have had alarmingly fat animals brought into their clinics at one time or another but they are not brought in because the client is concerned about their weight. Often the client simply does not recognise the obese condition as being detrimental to the animal, and will often be very reluctant to decrease the food intake. Fat animals are often seen as being cute and much loved. Thin, gaunt animals will provoke an immediate concern for them, even though statistics show many more animals suffer from complications of obesity than die from being too lean. Again, a lack of knowledge as to the needs of an animal, or even how an animal should look when fit, could be the problem.

Problems for Rabbits

More and more households are keeping rabbits as 'house rabbits'. Rabbits are very social animals and keeping them in the house can be a good experience if the needs of the rabbit are considered. Traditionally, rabbits are kept in a small cage in the garden. They are sometimes kept as a single rabbit and sometimes housed together in pairs. Studies done by Dr Shirley Seaman at the University of Edinburgh suggest that social contact is so important to rabbits that they will work hard in order to achieve even relatively short periods of visual and minimal tactile contact with another rabbit. Management of this contact is crucial.

Whether they are housed in a cage in the garden with other rabbits or kept in the household, rabbits prefer a living situation in which they have some control over the contact they have with other rabbits. Cages should be fitted with a bolthole that they can use if startled, or to avoid visual contact with other rabbits. Rabbits kept in adjacent cages could have some of the shared wall as a wire mesh in order to facilitate eye contact. In the house, rabbits should have a small cat carrier or box opening to the front in order to 'bolt' if frightened or wishing to avoid contact with other animals. Options in social contact make a rabbit feel secure.

Training and Welfare

Some people enjoy competing with animals and most people would suggest that competing with dogs is mutually beneficial to the dog and the people who do the training. This assumes that the trainer uses training techniques based on a thorough knowledge of the way that dogs learn, and has a clear understanding of the dog's behaviour. Even with the correct use of positive reinforcement there can be problems, such as with the timing of the reward, the clarity of the signals and other factors that can cause problems for the dog in making the correct associations. Breakdowns in training can lead to inappropriate use of aversive techniques and even the use of inhumane methods to force the animal to perform the behaviour. The pressure on trainers can be intensified by the need to produce a dog for competition.

All forms of manipulation should be continually reassessed in a training programme. The use of electric-shock collars should always be considered unethical but even standard choke collars can be cruel and dangerous. These collars are still in common use but should be considered unnecessary if more appropriate training methods are applied. There is plenty of evidence of the serious injuries that can be inflicted by these collars, not to mention the increased risk of promoting behavioural problems. Their use as a training 'aid' is certainly frowned upon by anyone who wishes to encourage welfare in the handling of dogs.

The use of any of these inhumane training devices or techniques by trainers is mainly based on the fact that they can successfully achieve the desired behaviour. However, we would argue that this is not a valid reason for their use. *The effectiveness of training and behavioural therapy should never be based solely on the success of the results. It should always be based on results achieved through the humane, ethical treatment of the animal.*

Problems with Breeding

Genetic manipulation of dogs and cats for the entertainment and enjoyment of human beings is another area for serious welfare concern. Dogs in particular have been selectively bred to enhance certain qualities required for hunting or fighting purposes. Some were selectively bred for a small stature, to be able to go down holes in the ground for prey, or fit more comfortably in a person's lap. Much of the more recent genetic manipulation that is popular in the dog is cosmetic. People want the dog to look a certain way, no matter how disastrous this is medically for the dogs involved.

Selective breeding for fashionable traits has serious welfare implications with dogs being bred with extreme deformities in order to fit a breed standard set by purebred dog enthusiasts. Of the breeds recognised by the Kennel Club, 148 breeds of purebred dogs collectively exhibit more than 300 genetic deformities and many of these breeds are affected by multiple inheritable problems. It could be argued that the breeders of these dogs deliberately select for characteristics that are known to cause medical problems for the animal.

Some of the more common breeds that are detrimentally affected by cosmetic breeding include the Cavalier King Charles Spaniel, Chow-Chow, the Basset Hound, the Dachshund, the Boxer, the Shar Pei, the Pug, the English Bull Terrier, the St Bernard and, perhaps the most damaged of all, the English Bulldog. Many of these deformities cause the animal great distress and/or pain throughout a sometimes-shortened life span. Since there is no other purpose for the selective breeding of these animals, this can only be attributed to human vanity.

A short list of the genetic disorders that are a direct

result of cosmetic breeding include: dental problems, injury-prone ears, breathing disorders, eye problems, shortened tear ducts, cataracts, inwardly turned eyelids, bone and joint failures, heart problems such as leaky valves and heart murmurs, cataracts, slipped kneecaps, spinal problems, hip dysplasia, skin diseases from skin folds, thyroid dysfunction, skull and brain deformities, and the list goes on and on. Professor David Morton, bioethicist and veterinarian at the University of Birmingham considers the issue to be one of 'integrity'. An animal has integrity if it is biologically fit in its natural state and capable of surviving in a range of environments.

Many breeds also suffer from behavioural problems that can be a direct consequence of breeding for function. Gundogs were originally bred to be very active dogs capable of, and requiring, a lot of exercise with enhanced retrieving abilities. This can cause problems of possessive aggression, or simply give rise to exercise requirements beyond the means of most people. Terriers were originally bred to be tough and tenacious hunters, not the lap dogs that many are now used as. Border Collies have proved to be exceptional at agility games, as they were originally bred to have high activity levels and heightened vigilance for aiding sheep herding. Now this breed is being bred for enhanced ability in agility competitions, raising questions as to how ethical this practice may be. The natural characteristics of the breed that were so valued in a functional context as sheepdogs, are now being honed through breeding so that they are more compulsive in their desire to chase a flyball.

Cats fare no better. Cosmetic breeding has created cats with no hair (Sphinx) and cats with such long, fine hair that they cannot properly groom themselves (Persian).

The Persian also has eye problems inherent with the aesthetically desirable short muzzle and nose. There are cats with foreshortened legs (Munchkin) which cannot jump or climb and very large cats (Maine Coon) that are prone to hip dysplasia. There are also cats with disorders that go along with the genetic deformities such as the tailless cat (Manx) and the potentially fatal gene attached to the Scottish Fold. Most of these animals are truly disabled and they cannot live natural, normal lives.

A new trend is the crossing of wild species of cats with domestic cats, to create an exotic animal for the pet trade. The Bengal cat and the Leopard Spotted Cat have a large percentage of wildcat in their gene pool. They are extremely active animals and they require a lot of attention and enrichment to keep them from being destructive. These cats prefer to be outdoors but, unfortunately, most of these cats are housebound as they are too valuable to allow out of the house. They have been bred mostly for their looks but little regard has been given to the needs of such an animal.

It is also popular to breed rabbits that would most likely not be able to live a natural life on their own. Some have large floppy ears dragging the ground. Others have fur that requires a daily brush from their human companion in order to keep it from matting.

All these animals have been bred for their looks and little regard has been given to their ability to participate in natural behaviour.

Does anyone care? The proposed new Animal Welfare Bill included a ban on docking puppies' tails; tighter controls of pet fairs and the exotic pet trade; a new criminal offence of 'likely to cause unnecessary suffering'; increased powers for police to investigate cruelty; longer sentences and more time for cases to be brought to court; and prosecution of breeders who produce genetically defective animals. Lobbying campaigns by breed associations and owners' clubs fiercely opposed every one of these proposals. These serious welfare issues, affecting millions of animals, are now in the hands of the Department of Environment, Food and Rural Affairs (DEFRA) and Parliament.

It could be argued that the deliberate breeding of cats and dogs should be eliminated altogether. Thousands of unwanted cats and dogs are abandoned, neglected or euthanased every day in Britain. Yet the breeding continues. Often the animals breed because their carers did not want to take responsibility for neutering them. So it is perhaps not so surprising that there is a segment of the population who would like to restrict or even ban people from keeping pets. Their reasoning is that they respect these animals too much to allow the practice of irresponsible pet ownership to continue.

Problems of Misconception

There are certainly welfare problems that arise due to owner ignorance as to the behavioural needs and physical requirements of their pet.

It is difficult to assess how much a dog will do without manipulation because, as a social animal, the dog usually wants to maintain the relationship with his owner. This does not necessarily mean that the dog would want to engage in a particular activity if left on his own.

A good example of misconception of an animal's needs is the walk that most dogs are *required* to take on a daily basis. It is commonly believed that a dog must be walked at least once a day to stay healthy. This is sometimes done with the purpose of calming the dog down through exercise. Sometimes these walks take the form of accompanying a person who is cycling or jogging. If the dog is mature and fit, these forms of exercise are probably beneficial and enjoyed, although it should always be recognised that dogs like a lot of slow, 'sniffy' walks in which to assess the world around them. It should also be pointed out that if the goal is to 'calm the dog down' by cycling/jogging, the dog probably just gets fitter. This has been known to make the dog fitter than the person and in doing so makes the dog more difficult to live with in the house.

Sometimes this perceived requirement to walk the dog is fraught with stress and anxiety for both the animal and the walker. The benefits of a walk can easily outweigh the costs in terms of stress, if the dog exhibits aggression or is over-fearful. It is often considered that dogs must be walked even if there is a big private garden available to the dog. Little thought is given to engaging the dog's brain by making the private environment more interesting. Perhaps more consideration should be given to games that can be played in the house or garden that provide sufficient exercise as well as mental stimulation. A dog with a behavioural problem involved with being in public could benefit from a new private regime. It would likely enhance the welfare of the dog, the walker and the general public if 'permission' were given to the individual to stop walking the dog in public until the problem is resolved or under better control. Confirming that it is okay to curtail walks and supplement with other games available can often be of benefit to all concerned.

A further problem is of old and infirm dogs who are required to continue the daily walk, even though it may no longer be enjoyable to them. A dog may continue to attempt this activity because it wishes to be with its human companion, or it may be a lifelong habit. Assumptions about the need and enjoyment of the activity should be assessed, and compromises made to benefit and accommodate the changing needs of the animal.

A potentially serious misconception is that animals have knowledge of wrongdoing. This is a welfare issue. Owners often state this because they feel that it is proven by the 'guilty' posture the animal appears to exhibit when confronted. While there is no scientific evidence that animals do not feel guilt, it is most likely that the posturing is done because the animal realises that the person is behaving in a strange way, even possibly aggressively.

Commonly misunderstood dog behaviour that can lead to an inappropriate owner interpretation includes the submissive grin that is often perceived to be a snarl and the play bow that some have interpreted as aggression. Clearly if we are going to have these animals in our lives, it is imperative that we educate ourselves on their social systems and behaviour.

The Kennel Environment

Dogs or cats may be placed in a kennel or cattery for short periods whilst the owner is on holiday, or in extreme cases, for longer periods if the animal is in quarantine. Alternatively a pet animal might find itself placed in a rehoming centre for various reasons. In all cases the animal is confined in an unfamiliar environment and exposed to fear-provoking stimuli such as strangers, noise or isolation. In the UK almost half a million cats are received in shelters each year, and in the US this number has been shown to be as high as 8 million. Changes in diet, social and physical environment all impose a challenge to the animal's physiology and psychology, and there are well documented cases of health problems that result from the animal being unable to meet the environmental challenge.

Cats may display behavioural inhibition, including failure to groom, feed or urinate; they may sleep for longer, hide more or show stereotypic licking or pacing. A more predictable environment where the animal can feel in control is often the best way to improve the welfare of these animals. Understanding the effect of the shelter environment on dogs can help to ensure that facilities for housing and management are improved. This in turn will lead to the development of fewer problem behaviours whilst the dog is awaiting rehoming. Currently 17.5 per cent of the overall shelter dog population are returned after rehoming. This in turn leads to further welfare problems where the dogs is separated from a new 'pack' leading to increased anxiety.

Symbiotic or Parasitic?

So is the keeping of animals as companions mutually beneficial or is it exploitation?

This question can lead to great confusion when discussing how we live with these animals. People keep dogs, cats and rabbits for various reasons. Most people keep these animals as 'members of the family', but this can be problematic for both animals and people, as pointed out above. Animals are often treated in a disrespectful manner, for example, as surrogate children, or as 'live toys' to be used by the family children. Both of these attitudes can lead to misinterpretation and anthropomorphisation of the animal's behaviour.

It is apparent that animals are often treated as objects rather than subjects. Sometimes described as the 'Disneyfication' of animals, this can lead to problems where animals have human characteristics projected on to them. In an extreme form of anthropomorphism, the animal is depicted as speaking human language, and its body is drawn in such a way as to resemble a human's. Of course most people recognise the difference between a cartoon figure and the real animal, but studies and experience have shown that this material can lead to a misinterpretation of the behaviour of animals by children and some adults. It can also lead to a limited knowledge of an animals' true behavioural and physical needs, and unrealistic expectations of an animal's behavioural responses. When these expectations are not met, people can be disappointed. They may respond with anger or withdrawal from the animal, leading inevitably, to a reduced state of welfare for the animal which may be expected to 'fit in', regardless of its needs, and this, in turn, will lead to behaviour problems, or it may even be discarded.

Finally we need to address the question of what motivates a person to keep an animal as a 'pet'. In some cases it has been suggested that the motivation may be the need to dominate and control members of another species. For others it could be that there is a substitution of animal affection for human affection. We need to consider whether we should be able to treat these animals as property to be bought, sold, neutered, fed and confined at our whim. And finally, we need to consider whether keeping animals as pets is a right, or a privilege.

14
Pet Ownership and the Law

Anne McBride

Introduction

WHAT IS LAW, WHERE DO COMPANION ANIMALS FIT IN? The subject of responsible ownership is often talked about when we consider the relationship between owners and their pets. This topic raises issues of moral responsibility for the animal's welfare and for that of people with whom the animal may have contact. What is less likely to be discussed is how such moral values and actions have been translated into law. That is, the professional being aware of the legal responsibilities of the owner, and ensuring the owner is also clear about these responsibilities. All too many owners fall foul of the law simply because they were not aware of how it applies to them, which is no defence in a court. In this case ignorance is not bliss.

As Broom succinctly stated in his preface to Radford's 2001 book on animal welfare law: the purpose of law 'is to promote cooperation among individuals and to discourage actions which harm individuals or society at large'. Laws are mechanisms by which moral actions can be promoted, and as we change our view of what is moral so this is reflected in law. This has been particularly evident with respect to the law relating to animal welfare, where changes over the last decades reflect our increasing knowledge of animal behaviour, emotion and cognition, and our need to discourage actions that harm individual animals.

Interestingly, such legal advances in animal welfare legislation have related to commercial aspects of ownership: farm animals, laboratory animals and dog breeders. Little has changed in the last century with respect to the welfare of animals kept as companions. Such a lack of change may, in part, be embedded in the cultural view that, as pet owners, we naturally will know what is best for our animal and thus 'do right by them', and thereby accommodate their welfare needs.

The concept and study of welfare is essentially a scientific issue, whose core meaning is summarised by the Farm Animal Welfare Council in 1993 in the Five Freedoms. These are that animals should have:

1. *Freedom from thirst, hunger and malnutrition* by ready access to fresh water and a diet to maintain full health and vigour.
2. *Freedom from discomfort* by providing a suitable environment including shelter and a comfortable resting area.
3. *Freedom from pain, injury and disease* by prevention or rapid diagnosis and treatment.
4. *Freedom to express normal behaviour* by providing sufficient space, proper facilities and company of the animal's own kind.
5. *Freedom from fear and distress* by ensuring condition and treatment which avoid mental suffering.

In order to be able to meet these freedoms an owner needs to have knowledge of the animal's ethology (its natural behaviour), nutritional and exercise requirements. In addition, an understanding of the appropriate socialisation, training and exposure to the environment that is needed to help prevent the animal suffering fear and distress.

Were it that merely being an owner was sufficient to provide us with the knowledge to 'know what is best', then we would see far fewer animals with behaviour problems; physical problems due to inappropriate breeding, such as hip dysplasia; less need for Pet Health advice on obesity in our pets and fewer teeth problems in rabbits due to inappropriate diet. These are just a few of the problems that beset companion animals. It is clear that merely the state of being a pet owner is insufficient to provide for an animal's welfare. Simply loving your pet is not enough: it is essential to have knowledge.

Currently, in the UK, the issue of welfare with respect to companion animals is not a legal concern as it is with other groups such as laboratory and farm animals. There is no 'duty of care' on the pet owner. 'Responsible pet ownership' as regards the animal's welfare as such is not a legal requirement. Interestingly, there is a means by which the welfare of companion animals could be much improved. This would be the adoption and implementation of the European Convention for the Protection of Pet Animals, which was developed by the Council of Europe in 1987. Whilst this has been ratified (made law) in several countries such as Belgium, Cyprus, Luxembourg and Portugal, successive UK Governments have declined to sign up to the Convention or to ensure that British law is in line with its provisions.

The Convention lays down requirements relating to all aspects of the animal's life, from selection of its parents (its breeding) to its death, including how it is kept and trained. For example, breeders would have a responsibility for the anatomical, physiological and behavioural characteristics that are likely to put at risk the health and welfare of either offspring or female parent. This would mean a responsibility for, amongst other things, adequate and appropriate experiences being provided to young animals during their early development. In addition, the Convention requires that there is the provision of information and education. It goes far beyond current UK legislation with respect to pet animal welfare.

Much of the current legislation relating to pet animals is concerned with the behaviour of dogs and how that impinges on society. There is specific legislation relating to horses as well, but this is outside the remit of this chapter, as is the law pertaining to the keeping of non-domesticated species such as snakes, large cat species, wolves and the like. The rest of the chapter will focus on the general legal responsibilities of the pet-, and in particular, dog-owner in the course of normal pet ownership. The text below attempts to translate the law into its underlying principles and general requirements, rather than spell out all the details of each. For more specific detail, readers are advised to consult the suggestions for further reading at the end.

As one gets to know the law relating to animals, it is apparent that much is ill-defined and there are many anomalies, even contradictions contained within it. This reflects the historical, almost organic nature of its development. Animal Law has grown rather than been planned. If one could compare it to the development of a city, it is more like the natural growth of London with its curved streets, blind alleys and almost impenetrable links and cross-streets, rather than the planned, simple, grid system of New York or Toronto. There is a recognised need for reform, but this is a mammoth task. However, the current consideration by the Government of animal welfare legislation is a step in the right direction.

Cruelty

The Protection of Animals Acts (1911, 1912) currently serve to protect dogs and other companion animals from cruelty. These are criminal acts and if the perpetrator is found guilty the courts can award fines, imprisonment, or a ban from keeping or owning animals either for a period, or permanently, and the removal, rehoming or destruction of the animal concerned. The acts cover many aspects of behaviour which may cause 'unnecessary suffering' to the animal. Unfortunately, the way the Acts are written, unnecessary suffering must be shown to have occurred (the exception being abandonment, where it is enough to show that unnecessary suffering is likely to be caused). The fact that someone's behaviour may 'be likely to' cause such suffering is not an offence. This means that an animal must be shown to have been actually suffering for an offence to be considered to have occurred. A minor change in the wording of this legislation would substantially improve the lives of many animals, allowing them to be removed from unsatisfactory circumstances before they were actually suffering. After all, surely prevention is better than cure.

Offences include:

- To cruelly kick, beat, over-ride, torture, infuriate, or terrify any animal
- To wantonly or unreasonably do, or omit to do any act which causes any unnecessary suffering.

Further provisions concern the way in which animals are transported or carried. It would probably be considered an offence to carry a rabbit by its ears, for example, or a dog in a crate that was too small. Fighting and baiting of any animal are also covered here, as are giving an animal any harmful drug or substance or operation. Operations on animals can only be carried out by qualified veterinary surgeons.

Of specific interest to dog owners is that under this Act it is an offence, though not one of cruelty, to use a dog as a draught animal – that is to pull a cart, sled, barrow or anything else on a public highway. So all of you with huskies or other breeds originally used for such a purpose, you can only use them for this on private ground.

Under the Abandonment of Animals Act (1960) which is linked to the above Protection of Animals Acts, it is an offence to abandon an animal, either temporarily or permanently, in circumstances which are likely to cause it unnecessary suffering – unless there is a reasonable excuse. This is an offence of cruelty and thus the sanctions under the Protections of Animals Acts apply, namely that if the owner is convicted an animal can be confiscated, and any person found guilty can be disqualified from owning or keeping animals.

Can You Have a Pet?

It is not always possible to own a pet. Some housing providers, be they private or local authority, have restrictions on what type of pet, if any, are allowed to be kept in their accommodation. Indeed, this also applies to leaseholders. Whether you are considering buying or leasing a property, you should always check that there are no conditions attached to that property that would restrict the keeping of animals.

If you are applying for rented accommodation, do

ensure that you are aware of the landlord's pet policy and be honest about whether you currently own, or intend to own a pet. Ensure that you obtain written permission from your landlord if you want to keep a pet. Not doing so can lead to difficulties and acrimony and can even result in eviction.

Tenants should also be aware that they are usually liable for any damage caused to the property, its fittings and furniture. Be warned it is not just dogs that can cause damage. Cats, birds and rabbits can ruin furniture and wallpaper and rabbits and rodents are not averse to chewing their way through wiring, potentially causing a fire hazard. Snakes and other reptiles are not unknown to be escape artists, turning up in strange places such as the boiler room of the apartment block: find that is unlikely to be appreciated by the caretaker, landlord or other tenants.

Dogs Out and About

There are several laws that relate to dogs when they are out and about. The first is that they should be with someone. A dog which is roaming unaccompanied is basically straying and the local authority has the right to seize it and detain it for seven days or until the owner collects it. After the seven days the animal can be rehomed, sold or humanely destroyed. If the owner is found, he or she is liable to pay for all expenses incurred because of the dog's detention. Should the owner turn up after the seven days to find the dog has been rehomed, for example, they have no recourse to get the animal back. As far as the law is concerned they have lost the ownership of that animal. This can be extremely distressing.

Thus it is in the owner's interest to ensure that the dog cannot escape from the property and, should it do so, that it is easily identified and its owner traceable. This is why it is a legal requirement for all dogs to wear a collar with the owner's name and address on it or on a tag attached to the collar. Any dog which does not have such a visible identification can be seized and treated as a stray. Having your dog tattooed or microchipped is an additional measure to help ensure that you and your pet can be quickly reunited. However, this does not negate the need for a visible collar and tag. Indeed, many dogs are reunited with their owner by other members of the public, simply because they were wearing a collar and tag. You yourself may have done just that. It is also worth noting that not all authorities have readers who can read all the varieties of microchip available, though this situation is improving. There is logic to the collar and tag rule. As an extra thought, dogs which wear a check chain should also be wearing a collar and tag. A check chain can slip off and the dog then is falling foul of the law and can be considered a stray.

Dogs which stray on to the road and cause an accident resulting in injury to a person or damage to property can land their owners with a substantial bill. The owner is liable for all costs associated with that damage, which could be many thousands of pounds. Likewise the owner of a dog which worries livestock is strictly liable for any damage caused, and the dog may well pay with its life, as livestock owners are legally entitled to shoot a dog seen worrying livestock. Some people think worrying means attacking and injuring livestock, but chasing is also considered worrying, as is a dog which is running among livestock so as to alarm them. Actual pursuit need not be proved. Certainly it is very worrying for the livestock, such as sheep, and may cause them to abort lambs or injure themselves in their fearful escape from what is a predator.

The bottom line then is to ensure that your dog cannot escape and that when you are out with your dog that it is kept on a lead on roads and lanes, and around livestock. Remember that all dogs are predators at heart and we cannot predict their every move, no matter how well they are trained. The sight of a cat or the shock of a car backfiring may cause your dog to run into the road and cause an accident. The simple expediency of a few feet of leash is well worth it.

Nuisance

The Environmental Protection Act 1990 intends to make our daily lives somewhat more pleasant. It relates to all sorts of nuisance, be it the noisy neighbours, street litter or nuisance caused by animals. It is an offence to keep any animal in such a place or manner as to cause a human health hazard, noise or other nuisance.

This obviously relates to keeping animals in appropriate accommodation and responsibly disposing of any waste so as to avoid any health hazard, or nuisance relating to the smell. It also has implications for animals which are left alone and cause a noise nuisance, be that a screeching parrot or barking dog. Often owners are unaware that their dog barks while they are out until they receive a visit from the local authority officer. This can cause a lot of distress that could be avoided.

As ever, it is worth taking preventive measures. Train your animal to be relaxed when it is left alone, advice on how to do this can be obtained from your veterinary surgeon, from the APBC and from leaflets produced by national rescue organisations. Ensure that the animal has had sufficient mental and physical exercise, and has appropriate things with which to occupy itself whilst you are out. Finally, ask your neighbours if there are any problems. Should your circumstances change, such as when the animal is left due to changes in your work pattern, ask them to let you know if any noise problems develop.

Fouling

Dog mess is unpleasant. It is smelly, difficult to get off your shoes and can pose a health hazard to people. Whilst it is not an offence to let your dog relieve itself out on a walk, it is an offence to fail to remove the faeces from 'designated land'. Designated land is all land open to the air to which the public have access, and which has been labelled as such by the local authority. Exceptions include land alongside a road with a speed limit of greater than forty miles per hour, agricultural land, woodland, marshland, moorland, heath and some common land. Under the Dogs (Fouling of the Land) Act 1996, if you do not pick up after your dog you are liable to pay a fixed fine of up to £1,000: a hefty price. My advice is to make sure you always have a supply of plastic bags in your pocket and that you dispose of used bags in the bin.

Where Can You Walk your Dog?

Different areas will have different restrictions on when and where dogs can be walked. These are enacted in local byelaws; details will be available from the local Council office. Some may prohibit dogs from particular areas, such as children's playgrounds. Such prohibitions may be applicable throughout the year; others may be seasonal, for example, those referring to beaches or fishing lakes. Other byelaws may state that in certain areas dogs must be kept on a lead.

Taking Your Pet Abroad

Under the Pet Travel Scheme, it is now possible to take your pet abroad, for example, if you go on holiday. Do consider if it is really appropriate to do so, especially if you are planning to go somewhere hot. Your pet may not appreciate the heat of the sun as much as you do.

If you do wish to take your pet abroad, you will need to plan several months ahead. Your pet will need to be microchipped and vaccinated against rabies. Blood tests will need to be taken to ensure that the vaccine has taken and the animal is protected against rabies. Depending on your destination, you may need to obtain import and/or export certificates. On return to the UK you will need to show that your animal has been treated for ticks and worms through the production of an appropriate veterinary certificate. There are several parasites that are common to other countries, but are rare or not found in the UK.

The Pet Travel Scheme is frequently updated with respect to the countries included and the species of animal to which it applies. You are advised to contact your veterinary surgeon for further advice. You may also wish to look at the Department of Farming and Rural Affairs (DEFRA) website which will provide up-to-date information.

You should also be aware that, whilst you are in another country, you will be subject to their animal legislation. For example, in some countries it is a legal requirement to have certain breeds of dog, or even all dogs, muzzled when in a public place. You need to make sure that you know what laws will apply to you and your pet.

Dangerous Dogs

There are two acts of law that relate to dangerous dogs. These are the Dogs Act 1871 (as amended by the Dangerous Dogs Act 1989) and the Dangerous Dogs Act 1991 (Amendment Act 1996). These are intended to ensure that the public is not at risk from dogs. It is a common misperception that these acts only relate to certain breeds, such as the Pit Bull Terrier. This perception relates to a series of well publicised dog attacks in the 1980s that stimulated the Government into formulating the 1991 Act. Whilst there are specific provisions for some breeds in the Dangerous Dogs Act, in fact both Acts refer to all dogs, be they toys, giant breeds, crossbreeds or mongrels. The Yorkshire Terrier or Papillon come under this legislation just as much as the German Shepherd, Labrador or Rottweiler.

The 1871 Act is a civil act, whereas offences under the Dangerous Dogs Act are criminal offences. The 1871 Act has other provisions that are thought by many to reflect a more considered approach, as opposed to the seemingly 'knee-jerk' reaction that resulted in the 1991 Act. However, rightly or wrongly, it is more common now for a case of a dog which is considered dangerous to be brought under the Dangerous Dogs Act, though the 1871 Act is still active and can be used.

Under the Dangerous Dogs Act it is a criminal offence to have 'a dog dangerously out of control in a public place'. Dangerously out of control means that there is reasonable cause to believe that a dog might injure somebody, whether or not it actually does so. So, for example, if your dog runs up to a member of the public and bounces around them barking, even jumping up, that person may have reasonable apprehension, or cause to believe that he or she may be injured and so bring a case against you.

Being considered as 'dangerously out of control' does not only relate to the dog's attitude to humans. The courts have also held that a dog can be dangerously out of control where it attacked another dog. There does not have to have been the involvement of any human in the fracas. This may not make sense from our knowledge of dog behaviour – as aggressive behaviour between two dogs can have various causes and can have no similarity to a dog's behaviour towards people, and vice versa. However, from the court's point of view a dog involved in a dog fight has been deemed a reasonable cause to believe that a person might be injured by that dog and thus deem that dog as dangerous.

If no injury to a person has occurred, then the courts may invoke a control order. This could involve your dog being restricted to being on a lead, muzzled or banned from being in certain areas. Should an injury occur, no matter how slight, the dog may be destroyed. In either case you will have a criminal conviction. It is important to be aware that injury in this case does not just mean that the dog has bitten. It may have scratched someone when jumping up, or caused them to fall over or off their bicycle.

This may sound rather harsh and there are views that this Act has drawbacks compared to the earlier 1871 Act. However, we should always consider the wider social context of all legislation. Dogs, like cars, unless controlled, can cause damage. Though of course owners as well as legislators need to remember that, unlike cars, dogs are animals and thus their behaviour is not totally predictable.

Again, prevention is better than cure

Adequate and appropriate socialisation is essential. Socialisation to people of all kinds – of different ages, mobility and colour, as well as to other dogs, of all shapes and sizes – means that a dog is unlikely to cause damage to a person through fear, or be considered to be dangerous by getting into dog fights. Appropriate training of basic manners on how to approach people is also important. Teach your dog to wait for permission to approach people and to sit when it does so. This can be started in puppyhood and means that your dog will be an ambassador for dogdom, rather than an advert for the less desirable qualities of the canine race.

Breeding of Dogs

In 1999 the Breeding and Sale of Dogs (Welfare) Act was invoked. This amends the Breeding of Dogs Act, which remains in force, subject to the amendments introduced by the 1999 Act. As its name implies, this is intended to improve the welfare of breeding dogs and their puppies. If you breed dogs as a business activity, you are now required to obtain a licence from your local authority. If you breed five litters a year, you are automatically deemed to come under this Act. If you breed fewer than five litters a year, then it is open to the local authority to decide whether or not you are breeding dogs as a business. Should you be required to obtain a licence, your premises will be inspected by a veterinary surgeon before a licence is granted and will be inspected regularly thereafter. The Act also has specific restrictions on the breeding and/or sale of puppies. Anyone found to not have a licence will be fined and may face a prison sentence.

You may wish to breed from your dog and do not come under the specifications of the licence. Do think carefully, however. No matter whether you breed only one litter or several, homes are not always easy to find and even then it is not guaranteed that a puppy, which is very cute when young, will still be wanted when it is older. The number of animals given to rescue societies because they are no longer wanted is testimony to this.

Insurance

Unlike car ownership, there is no legal requirement to have your animal covered by Third Party Liability insurance. However, it is advisable to do so. It will normally cover you against damage caused by your animal to other people or property. It is often included as part of household insurance policies, and it is worth checking as to whether that is true of your policy. It is also available from pet health insurance companies and, in the case of dogs, can be obtained separately from Pro Dogs.

It is also worth considering insuring the health of your pet. There is no National Health system for pets and veterinary fees can soon mount up. There are many policies available and it is worth shopping around to find the one most suited to you and your pet.

As with all insurance cover, do read the small print and ensure that the policy covers all the relevant issues.

Final Thoughts

This chapter has only briefly covered the legislation that is most commonly enforced with respect to pet ownership. There are other laws that have not been discussed, such as those relating to dogs and game or the keeping of dogs as guard animals. The law is not a static institution. Rather, it does change and the latter years of the twentieth century and start of the twenty-first have seen many changes occur or be placed under consideration. The proposed Animal Welfare Bill and the Hunting with Dogs Bill both have implications for the pet owner. Just what these will be is still under discussion.

Pet owners have responsibilities, some of which are enshrined in law, to their pet and to the wider community. The quality of life of the animals in our care is ours and ours alone. They are totally dependent on us for their physical and mental well-being, and we should consider whether or not we can meet this responsibility before we acquire any pet.

With respect to the wider community, both human and animal, we need to ensure that our animals are not likely to cause harm or nuisance. Good breeding and early experience will set the animal on the right path. However, this needs to be followed up with appropriate experiences and training. In particular, dog owners need to be aware of teaching their dog to respond quickly and reliably to basic commands such as 'come', 'stay', 'sit', 'down', 'heel' and 'leave'. Basic manners of not jumping up or running towards people are not only courteous, after all not everyone likes dogs, they will also prevent any misconceptions about your dog's intentions. Owners also need to ensure that their dog interacts well with other dogs. If there are any problems, take appropriate measures, which may be keeping it on a lead or muzzled, *and* seek help. Remember the earlier help is sought the better: problem behaviour often worsens if left.

If you are not planning to breed from your pet, be it

dog, cat or rabbit, then it is advisable to have it neutered. This will prevent unwanted litters. In the case of male dogs, neutering reduces the chances of the dog being involved in fights or roaming and possibly being picked up as a stray.

Responsible ownership is an aspect of our general social responsibilities. The law merely serves to underline this basic concept. We need to remember that both our pets and ourselves are part of a wider community. Taking appropriate action, exercising foresight and common sense, will allow us all to live in harmony and for our pets to be considered as valued members of society.

References and Useful Reading

1. *Evolutionary Aspects of Canine Behaviour*

REFERENCES

Goodwin, D., Bradshaw, J. W. S. & Wickens, S. M. (1997). 'Paedomorphosis affects agonistic visual signals of domestic dogs', *Animal Behaviour* 53: 297–304

Vila, C., Maldonado, J. E. & Wayne, R, K. (1999). 'Phylogenetic relationships, evolution, and diversity of the domestic dog', *J. Heredity* 90(1): 71–77.

USEFUL READING

Clutton-Brock, J. (1995) 'Origins of the dog: domestication and early history', in *The Domestic Dog*, Serpell, J. (ed.), Cambridge University Press.

Davis, S. J. M. & Valla, F. R (1978). 'Evidence for the domestication of the dog, 12,000 years ago in Israel', *Nature* 267: 608–10.

Ginsburg, B. E. (1987). 'The wolf pack as a socio-genetic unit', in *Man & Wolf*, Frank, H. (ed.), Dr W. Junk Publishers, Dortrecht, Nethelands.

McFarland, D. (1999). *Animal Behaviour*, 3rd edn. Longman.

Mech, L.D, (1997). *Ten Years With the Pack*, Swan Hill Press, Shrewsbury.

Mech, L.D, (1979). *The Wolf: The Ecology and Behaviour of an Endangered Species*. University of Minnesota Press.

Savolainen, P., Ya-ping Zhang, Jing Luo, Lundeberg, J., Leitner, T. (2002). 'Genetic evidence for an East Asian origin of domestic dogs', *Science* 298: 1610–13.

Vila, C., Savolainen, P., Maldonado, J.E. amorim, I.R., Rice, J.E., Honeycutt, R.L., Crandall, K.A., Luneberg, J. and Wayne, R.K. (1997). ('Multiple and ancient origins of the domestic dog', *Science*, vol. 276:1687–9.

Woolpy, J. H (1968). 'The social organisation of wolves', *Nat. Hist.* 77 (5): 46–55.

2. The Foundations of Canine Behaviour

REFERENCES

Appleby, D.L., Bradshaw, J.W.S. and Casey, R.A. (2002). 'Relationship between aggreessive and avoidance behaviour by dogs and their experience in the first six months of life', *Veterinary Record*, 150: 434–8.

Cairns, R.B. & Werboff, J. (1967). 'Behaviour Development in the Dog: An Interspecific Analysis', *Science*, 158:1070–2.

Elliot, O. and Scott, J.P. (1961). 'The development of emotional distress reactions to separation in puppies', *J. Genet. Psychol.*, 99:3–22.

Fox, M.W. (1971). *Integrative development of Brain and Behaviour in the Dog*, University of Chicago Press.

Fox, M.W. and Stelzner, D. (1996) 'Behavioural effects of differential early experience in the dog', *Animal Behaviour*, 14, 273–81.

Freedman, D.G., King, J.A. and Elliot, O. (1961). 'Critical period in the social development of dogs', *Science*, 133:1016–17.

Pluijmakers, J., Appleby, D.L., Bradshaw, J.W.S. (2003). 'Sensitive Periods in the Development of Behavioural Organization in the Dog and the Role of Emotional Homeostasis', *Proceedings of the 4th International Behavioural Meeting*, Caloundra, Australia. University of Sydney.

USEFUL READING

Freeman, D. (1991) *Barking up the Right Tree*, Ringpress Books.

Gray, J. (1971) *The Psychology of Fear and Stress*, Cambridge University Press.

Panksepp, J. (1998) *Affective Neuroscience: The Foundation of Human and Animal Emotions*, Oxford University Press.

Scott, J.P. and Fuller, J.L. (1965) *Genetics and the Social Behaviour of the Dog*, University of Chicago Press.

Vincent, J.D. (1986) *Biologie des Passions*, Odile Jacob.

3. *Puppy Classes –a League of their Own*

USEFUL READING

Appleby, D. (1993). *The Behaviour of Dogs and Cats*, Stanley Paul.

Appleby, D. (1997). *Ain't Misbehavin'*, Broadcast Books.

Campbell, W.E. (1985), *Behavior Problems in Dogs*, American Veterinary Publications, Inc.

Dunbar, I. (1996), *Doctor Dunbar's Good Little Dog Book*, James and Kenneth Publishers.

Fogle, B. (1990), *The Dog's Mind*, Penguin.

Klinghammer, E., *Wolf Park*, Indiana, USA Personal communication to author.

Lorenz, K. (1955), *Man Meets Dog*, Methuen and Co. Ltd.

McCune, S., McPherson J.A., and Bradshaw, J.W.S. 1995, *The Waltham Book of Human Interaction – Benefits and responsibilities of pet ownership*, Pergamon.

Peachey, E. (1992), *Running Puppy Classes*, self published

Peachey, E. (2003), *The 'Good Puppy!' booklet*, self published

Peachey, E. (1998), *The 'Good Dog!' booklet*, self published

Peachey, E. (2001), *Dog behaviour*, Paragon.

Pluijmakers, J., Appleby, D.L and Bradshaw, J.W.S. (2003). 'Sensitive periods in the Development of Behavioural Organisation and the Role of Emotional Homeostasis', 4th International Behavioural Meeting, Caloundra, Australia.

Serpell, J. and Jagoe, J.A., 1995, *The Domestic Dog*, Cambridge University Press.

5. *The Importance of Positive Reinforcement*

REFERENCES

APBC (2002), *Annual Review of Cases.*

Azrin, N.H., Holz, W.C. & Hake, D.F. (1963), 'Fixed-ratio punishment', *Journal of the Experimental Analysis of Behaviour*, 6, 141–8.

Ben-Michael, J. (1997), 'Emotional Experiences of Dog Owners Facing Disciplining Situations With Dogs', *Proceedings of the First International Conference on Veterinary Behavioural Medicine*, United Federation for Animal Welfare.

Carlson, N.R. (1994), *Physiology of Behaviour*, Chapter 11, Emotion and Stress, Allyn & Bacon

Cheyne, J.A. (1969), 'Punishment and reasoning in the development of self-control', Paper presented at the R.H. Walters Memorial Symposium at the Biennial Meeting of the Society for Research in Child Development.

Dweck, C.S. & Licht, B.G. (1980), 'Learned helplessness and intellectual achievement', in J Garber & M.E.P. Seligman (Eds.), *Human Helplessness: Theory and applications*, New York, Academic Press.

Gross, R.D. (1992), *Psychology – The science of mind and behaviour*, 2nd edn., Hodder & Stoughton.

Jones-Baade, R.E. (2001), 'The influence of the owner on the development of aggressive behaviour in dogs', Proceedings of the Third International Congress on Veterinary Behavioural Medicine, United Federation For Animal Welfare.

Sears, R.R., Maccoby, E.E. and Levin, H. (1957), *Patterns of child rearing*, Evanston, IL: Row, Peterson.

Solomon, R.L., Turner, L.H. and Lessac, M.S. (1968), 'Some effects of delay of punishment on resistance to temptation in dogs', *Journal of Personality and Social Psychology*, 8, 233–8.

Pavlov, I.P. (1927), *Conditioned Reflexes*, Oxford University Press.

USEFUL READING

Appleby, D. (1997), *Ain't Misbehavin'*, Broadcast Books.

Lieberman, D.A. (1993), *Learning: Behavior and Cognition*, Brooks/Cole Publishing Company.

Kershaw, E. (1998), *Go Click*, Pet Behaviour Centre.

6. *Scent Communication in Dogs*

REFERENCES

Ackerman L. (1996), *Dog Behavior and Training Veterinary Advice for Owners*, T.F.H Publications.

Appleby D. L. (1990), *The Good Behaviour Guide*, Pet Behaviour Centre.

Appleby D. L. (1997), *Ain't Misbehavin'. A Good Behaviour Book Guide for Family Dogs*, Broadcast Books.

O Farrell V. (1992), *Manual of Canine Behaviour*, 2nd edn. BSAVA.

Overall K. (1997), *Clinical Behavioural Medicine for Small Animals*. Moseby Publishing.

Price C. (2001), *Understanding the Rescue Dog*, Broadcast Books.
Serpell, J (1995), *The Domestic Dog: Its evolution, behaviour and interactions with people*, Cambridge University Press.

7. The Relationship between Emotions and Canine Behaviour Problems

REFERENCES

Appleby D., Pluijmakers J. (2003), 'Separation anxiety in dogs: the function of homeostasis in its development and treatment', *The Veterinary Clinics of North America,* 33, 321-344
Carlson N.R. (1998), *Physiology of Behaviour*. Sixth edition, Allyn and Bacon.
Frijda N.H. (1988), *De Emoties. Een overzicht van onderzoek en theorie*. Uitgeverij Bert Bakker.
Jones-Baade R. & McBride A, (1999), *A Biological Model For Aggression in dogs*, Mondial d' Etholie, Lyon, 28–41.
O'Farrell V, (1995), 'Effects of Owner Personality and Attitudes on Dog Behaviour', in Serpell, J.A. (ed.), *The Domestic Dog; Its Evolution, Behaviour and Interactions With People*. Cambridge University Press.
Pluijmakers J., Appleby D. L. and Bradshaw J.W.S. (2003), 'Sensitive Periods in the Development of Behavioural Organization in the Dog and the Role of Emotional Homeostasis', *Proceedings of the 4th International Veterinary Behavioural Meeting*, No: 32, 18th–20th August 2003, Caloundra, Australia, 119–26.
Rolls E.T. (1990), 'A Theory of Emotion, and its Application to Understanding the Neural Basis of Emotion', *Cognition and Emotion*, 4 (3), 161-190
Solomon R.L., Corbit J.D. (1974), 'An Opponent-process Theory of Emotion: I. Temperal Dynamics of Affect', *Psychological Review*, 81 (2), 119–45.

8. Feline Behaviour Problems – The Influence of Natural Behaviour

USEFUL READING

Bradshaw, J. (1988), *The Behaviour of the Domestic Cat*, CAB International.

Turner, Dennis and Bateson, Patrick, (eds). (1998). *The Domestic Cat – The Biology of its Behaviour*, Cambridge University Press.
Heath, Sarah, (1993), *Why Does My Cat...?* Souvenir Press.
Lansberg, G., Hunthausen, W. and Ackerman, L. (1997). *Handbook of Behaviour Problems of the Dog and Cat*. Butterworth Heinemann.
Robinson, Ian (ed.) (1995), *The Waltham Book of Human Animal Interaction*, Pergamon Press.
Overall, Karen (1997) *Clinical Behavioral Medicine for Small Animals*, Mosby

9. Behaviour Problems in the Domestic Rabbit

USEFUL READING

Lockley, R.M. (1980) *The Private Life of a Rabbit*, Book Club Associates.
Magnus, Emma (2002) *How to Have a Relaxed Rabbit*, Pet Behaviour Centre.
McBride, Anne (1988) *Rabbits and Hares*, Whittet Books.
McBride, Anne (2000) *Why Does My Rabbit?*, Souvenir Press.

10. Rage syndrome – in our dogs' minds?

USEFUL READING

Colter, S. B. (1989). 'Complex partial seizures: Behavioral epilepsy', in *Problems in Veterinary Medicine, Vol 1(4): Epilepsy*, 619–627, Richard J. Indrieri, ed, J. B. Lippincott.
Delgao-Escueta, A. V., Mattsonb, R. H., Kong, L. *et al.* (1981). 'The nature of aggression during epileptic seizures', *New England Journal of Medicine,* 305: 711–16.
Dodman, N. H., Knowles, K. E., Shuster, L., Moon-Fanelli, A. A., Tidwell, A. S. and Keen, C. L. (1996). 'Behavioral changes associated with suspected complex partial seizures in Bull Terriers', *Journal of the American Veterinary Medical Association*, 208(5): 688–91.
Dodman, N. H., Miczek, K. A., Knowles, K., Thalhammer, J. G. and Shuster, L. (1992). 'Phenobarbital-responsive episodic dyscontrol (rage) in dogs'. *Journal of the American Veterinary Medical Association* 201(10): 1580–3.
Fogle, B. (1990). *The Dog's Mind*. Pelham Books.
Hart, B. L. and Hart, L. A. (1985). *Canine and Feline Behavioral*

Therapy. Lea & Febiger.

Kelley, P. & Lewis, S. (eds) (1989). eds. *The Brownings, Correspondence. Vol 11. July 1845 – January 1846, Vol 12. January 1846 – May 1846*. Wedgestone Press.

Lloyd Carey, J. (1992). *Cocker Spaniels – An Owner's Companion*. Crowood Press.

Marsh, L. and Krauss, G. L. (2000). 'Aggression and violence in patients with epilepsy'. *Epilepsy and Behavior* 1: 160–8.

Mugford, R. A. (1984). 'Aggressive behaviour in the English Cocker Spaniel', *Veterinary Annual* 24: 310–14.

Neville, P. (1991). *Do Dogs Need Shrinks?* Sidgwick & Jackson.

Podberscek, A. L. and Serpell, J. A. (1996). 'The English Cocker Spaniel: preliminary findings on aggressive behaviour', *Applied Animal Behaviour Science* 47: 75–89.

Thomas, W. B. (2000). 'Idiopathic epilepsy in dogs', *Veterinary Clinics of North America: Small Animal Practice* 30(1): 183–206.

12. Behavioural Problems in Ageing Pets

USEFUL READING

Cantanzaro, T.E. (1999). 'Care of the Ageing Pet', in Jevring, C. and Cantanzaro, T.E. (eds), *Healthcare of the Well Pet*, W.B. Saunders pp. 49–65.

Cummings, B.J., Su, J., Cotman, C., White, R., Russel, M. (1993) 'Beta Amyloid accumulation in aged canine brain: A model of early plaques formation in Alzheimer"s disease', *Neurobiology of Ageing*, 14: 547–60

Cummings, B.J., Head, E., Afagh, A.J., Milgram, N.W., Cotman, C.W. (1996a), 'Beta Amyloid accumulation correlates with cognitive dysfunction in the aged canine', *Neurobiology of Learning and Memory*, 66: 11–23.

Cummings, B.J., Head, E., Ruehl, W., Milgram, N,W., Cotman, C.W., (1996b) 'The canine as an animal model of human aging and dementia', *Neurobiology of Ageing*, 17: 259–68.

Dodman, N.H. (1998). 'Geriatric Behavior Problems', Dodman, N.H. and Shuster, L. (eds), in *Psychopharmacology of Animal Behavior Disorders*, Blackwell Science, pp.279–82.

Head, E. (2001), 'Brain aging in dogs: Parallels with human brain aging and Alzheimer's disease', *Vet. Ther.* 2(3) 247–60.

Kitani, K., Kanai, S., Ivy, G.O., Carrillo, M.C. (1998), 'Assessing the effects of deprenyl on longevity and antioxidant defences in different animal models', *Ann. NY Acad. Sci.* 854; 291–306

Landsberg, G., Hunthausen, W. and Ackerman, L. (1997) *Handbook of Behaviour Problems of the Dog and Cat*, Butterworth Heinemann, pp. 185–94.

Landsberg, G.M., Moffat, K. and Head, E, (2003) 'Prevalence, clinical signs and treatment options for cognitive dysfunction in cats', *Proceedings of the fourth International Veterinary Behaviour Meeting Caloundra Australia*, pp.77–83.

Milgram, N.W., Head, E., Weiner, E., Thomas, E. (1994) 'Cognitive functions and aging in dogs: Acquisition of nonspatial visual tasks', *Behav. Neurosci.* 108; 57–68.

Milgram, N.W., Estrada, J., Ikeda-Douglas, C., *et al.* (2000) 'Landmark discrimination learning in aged dogs is improved by treatment with an antioxidant enriched diet', *Abstr. Soc. Neurosci.* 193: 9.

Milgam, N.W., Head, E., Cotman, C.W., Muggenburg, B., Zicker, S.C. (2001) 'Age dependent cognitive dysfunction in canines: Dietary intervention', in Overall, K.L., Mills, D.S., Heath, S.E., Horwitz, D. (eds), *Proceedings of the 3rd International Congress on Veterinary Behavioural Medicine* UFAW, pp.53–7.

Nikolov, R., Dikova, M., Nikolova, T., Nerobkova, L., Garibova, T. (1987) 'Cerebroprotective effect of nicergoline and interference with the anti-hypoxic effect of prostacyclin', *Methods and Findings of Experimental Clinical Pharmacology* 9: 479–84.

Pageat, P. (2001). 'Description, clinical and histological validation of the ARCAD score (evaluation of Age Related Cognitive and Affective Disorders)' in *Proceedings of the 3rd International Congress on Veterinary Behavioural Medicine*, pp.83–8.

Penalligon, J. (1997) 'The use of nicergoline in the reversal of behaviour changes due to ageing in dogs: a multi-centre clinical field', Mills, D.S., Heath, S.E., Harrington, L.J. (eds), in: *Proceedings of the 1st International Conference on Veterinary Behavioural Medicine*. UFAW, Potters Bar pp 37–41.

Postal, J.M., Van Gool, F.V. and Consalvi, P.J. (1994) 'Use of Nicergoline in the ageing dog', in *Proceedings of the 19th WSAVA Congress Durban South Africa*, p.781.

Neilson, J.C., Hart, B.L., Cliff, K.D. and Ruehl, W.W. (2001)

'Prevalance of behavioural changes associated with age related cognitive impairement in dogs', *JAVMA*, 218 (11): 1787–91.

Ruehl, W.W., DePaoli, A. and Bruyetter, D. (1994). 'Pretreatment characterization of behavioral and cognitive problems in elderly dogs', *J. Vet. Int. Med.* 8:178.

Ruehl, W.W. and Hart, B.L. (1998) 'Canine cognitive dysfunction', Dodman, N.H. and Shuster, L. in *Psychopharmacology of Animal Behavior Disorders*, Blackwell Science, pp.283–304.

13. *Issues of Companion Animal Welfare*

USEFUL READING

Radford, M. (2001) *Animal Welfare Law in Britain: Regulation and Responsibility*, Oxford University Press.

Wise, Steven, M. (2000) *Rattling the Cage: Toward Legal Rights for Animals*, Perseus Publishing.

Bekoff, Marc (1998) *Encyclopaedia of Animal Rights and Animal Welfare*, Fitzroy Dearborn Publishers.

Serpell, James (1996) *In the Company of Animals*, Cambridge University Press.

Waran, Natalie (2002) *The Welfare of Horses*, Kluwer Academic Publishers.

14. *Companion Animals and the Law*

USEFUL READING

Radford, M (2001). *Animal Welfare Law in Britain*, Oxford University Press.

Sandys-Winsch, G. (1993) *Dog Law Handbook*, Shaw and Sons. This is updated as and when appropriate and is useful for the professional who needs to know about the law in more depth.

OTHER SOURCES OF INFORMATION

Citizens Advice Bureau

Local Authority

Website of Department of Farming and Rural Affairs (DEFRA) at www.defra.gov.uk

Website of Her Majesty's Stationery Office (HMSO) at www.hmso.gov.uk

The Association of Pet Behaviour Counsellors (APBC)

The Association of Pet Behaviour Counsellors was founded in 1989 to promote and develop the profession of pet behaviour counselling and standardise the service provided. It aims to provide a network of specialist counsellors throughout the UK and internationally to whom veterinarians can confidently refer clients. In order for veterinarians to feel confident about referrals, Full Members of the APBC have to have the highest professional standards, knowledge and expertise. For this reason, a rigorous selection procedure is in place to assess applicants wishing to join the Association to ensure they meet the Association's stringent criteria.

Whilst studying for relevant academic qualifications and/or gaining practical experience, there are two levels of membership available, Student and Provisional. Full details of membership are available on the APBC's website, given below. The APBC also has a book and product mail order department.

The APBC enables the membership to act as a forum for the exchange of information and ideas in the rapidly growing and exciting field of animal behaviour therapy. Members receive a quarterly newsletter to which they are welcome to submit articles and are kept informed of new innovations, techniques, research and products which may help them with their work. They can take part in workshops and educational days, run exclusively for members on specialist topics. With a growing number of overseas members, a worldwide network enables contacts and exchange of information internationally as well as in the UK. An exclusive e-mail server for the membership ensures a rapid response to a request for information. As well as being able to discuss problematic cases within the membership, the APBC also has a team of specialist advisors from whom assistance can be sought. This includes many highly respected academics who are leaders in their fields. The APBC holds regular conferences and seminars around the country that are open to non-members.

APBC
PO Box 46
Worcester
WR8 9YS
Tel: +44(0) 1386 751151
Fax: +44 (0) 1386 750743
Website: www.apbc.org.uk
email: apbc@petbcent.demon.co.uk

Useful Contacts

The Companion Animal Behaviour Therapy Study Group (CABTSG)
Affiliated to the British Small Animal Veterinary Association, CABTSG is open to veterinary surgeons, veterinary nurses and others practising in the field of pet behaviour. It exists to facilitate the exchange of information, experiences and ideas between its members and to this end holds an annual study day and publishes a newsletter for its members.
www.cabtsg.org.uk

European Society of Veterinary Clinical Ethology (ESVCE)
The ESCVE is a non-profit making organisation which aims to promote and support scientific progress in veterinary behaviour medicine and comparative clinical ethology. www.esvce.org

Association for the Study of Animal Behaviour (ASAB)
ASAB was founded in 1936 to promote the study of animal behaviour, and membership is open to all who share this interest. There are now approximately 2,000 members, mostly drawn from Britain and Europe. Many members are professional biologists who work in universities, research

institutes or schools. ASAB launched a registration scheme for practising pet behaviour counsellors in 2003, believing that certification is the means by which it can be demonstrated to the public and to other professions, such as veterinarians, that certain individuals meet the minimum standards of education, experience and ethics required of a professional clinical animal behaviourist. www.societies.ncl.ac.uk/asab

List of Contributors

David Appleby MSc, CCAB

David has been in practice as a pet behaviour counsellor since 1986 was a founder member of the APBC in 1989. He is based at the Pet Behaviour Centre® in Worcestershire, UK, in addition to which he holds veterinary referral clinics in Birmingham, Derby, Leicester, Northampton, Wolverhampton and Nottingham. David is one of the first pet behaviour counsellors to be certificated by the Association for the Study of Animal Behaviour, a new registration and certification scheme launched in 2003. He is the visiting behaviour counsellor at Queens Veterinary School at Cambridge University. David is active within the profession generally and at the time of writing is Honorary Secretary for the Association of Pet Behaviour Counsellors and membership secretary for the British Small Animal Veterinary Association (BSAVA) affiliated Companion Animal Behaviour Therapy Study Group. He is a frequent lecturer at home and overseas, contributes to a number of journals, and lectures on the postgraduate diploma/MSc course for pet behaviour counselling at Southampton University. He will be familiar to many in the UK as the author of a series of client-orientated behaviour booklets stocked by veterinary practices and *Ain't Misbehavin' – a good behaviour guide for family dogs.*

Jan Hoole PhD DipCABC
Jan is based in Shropshire where she runs her behaviour practice, Ashley Pet Behaviour Centre, part of the Pet Behaviour Centre® group of practices. Since gaining her PhD Jan has taught part time at the University of Central England, and both full and part-time at Keele University. Subjects taught include Molecular Genetics, Evolution, Physiology and Pathology, Immunology and Animal Behaviour, to name but a few. Jan also presents guest lectures at Harper Adams University and the Merseyside Police Dog Training Unit.

Jolanda Pluijmakers DipCABC
Jolanda holds a post-graduate diploma in Companion Animal Counselling. She runs clinics on referral of vets in her practice, the Animal Behaviour Clinic, and a number of veterinary clinics in the Netherlands. She is studying for a PhD at the faculty of Veterinary Medicine at Bristol University. Her thesis will be on the behavioural development of the dog, particularly its emotional development.

Erica Peachey BSc(Hons)
Erica graduated from Hull University with a BSc (Hons) in Psychology, specialising in Applied Animal Ethology. She then went to the Royal (Dick) School of Veterinary Studies, as a research associate, to investigate the behaviour effects of spaying bitches. As prevention is better than cure, Erica runs classes for pet dogs, aiming to teach owners and dogs to have a better understanding of each other, to avoid problems and to teach response to basic commands. As well as pet behaviour counselling she also offers individual training sessions and rehabilitation

training. Her special interests include puppy education and children and pets. Erica frequently acts as an expert witness in court cases relating to dogs and is a regular contributor to the written and spoken media.

Inga MacKellar MSc

Inga has been involved in fostering cats and dogs for many years for various animal charities, through which she gained an interest in behaviour. Despite coming from a marketing and advertising background she was fortunate to be given the opportunity to develop a new career as a mature student – which she eagerly took up. She has been practicing as a pet behaviour counsellor for six years, and holds an MSc in companion animal behaviour counselling from Southampton University.

Inga lives in rural Sussex, where she practises, along with her six dogs, ten cats and a very patient and supportive husband.

Rosie Barclay BSc(Hons)

Rosie is a companion animal behaviourist practising in Nottinghamshire, South Yorkshire, Lincolnshire and Humberside. She holds an BSc(Hons) in Zoology and M.Phil in Animal Behaviour and Welfare. She previously worked as a research assistant at Nottingham University, as an animal behaviour lecturer at Derby University and as a lecturer in veterinary science at Broxtowe College, Nottingham. She has also been an editorial assistant on the international scientific journal *Animal Behaviour*. Before attending university Rosie worked as a veterinary nurse for over six years.

Charlotte Nevison PhD

Charlotte has always been fascinated by the behaviour of animals – especially the strange things they do that we as mere humans find difficult to explain! Her Puli's, horses and rodents are quite used to 'Mum' closely observing and noting their toileting patterns and social sniffing. Why do they, and most other animals, seem to be so interested in smells – including those that we find revolting or social no-no's?

Sarah Heath BVSc MRCVS

Sarah qualified as veterinary surgeon in 1988 and spent four years in mixed general practice. She set up a behavioural medicine referral practice in 1992. Sarah is an honorary lecturer in the Department of Veterinary Clinical Sciences and Animal Husbandry at Liverpool University and recognised teacher in the Department of Clinical Veterinary Science at Bristol University. Sarah is also an external tutor on the postgraduate diploma in Companion Animal Behaviour Counselling at Southampton University. She is a member of the Association of Pet Behaviour Counsellors and at the time of writing is President of the European College of Veterinary Behavioural Medicine – Companion Animals, and the European Society of Veterinary Clinical Ethology.

E. Anne McBride PhD

Anne holds a BSc (Hons) degree in Psychology awarded by University College London in 1978. She was awarded her Doctorate in animal behaviour from the same institution in 1986 and in 1992 she obtained a Certificate in Conservation and Ecology from Birkbeck College, London. She is a Fellow of the Royal Society of Arts. Anne has been a practising animal behaviour therapist since 1987 and runs the Animal Behaviour Clinic at the University of Southampton. As well as being a member of the Association of Pet Behaviour Counsellors she is an honorary member of the UK Registry of Canine Behaviourists. She writes and lectures, nationally and internationally, on various aspects of animal behaviour and the human-animal bond. In 2003 she co-founded, with Emma Magnus and Georgie Hearne, the Rabbit Behaviour Advisory Group. Since 1991 Dr McBride was associated with the Anthrozoology Institute as a Research Fellow and was Deputy Director from 1994 to 1999. She is course director for the Postgraduate Diploma/MSc in Companion Animal Behaviour Counselling at Southampton University – the first British academically recognised qualification. She is an Honorary Teacher at the University of Bristol Veterinary School where she lectures on animal behaviour. Anne has a particular interest in Animal Law where it is pertinent to pet owners.

Emma Magnus MSc CCAB
One of the first people to be certificated by the Association for the Study of Animal Behaviour, Emma has been practising as a pet behaviour counsellor since 1995. She has run monthly referral clinics at veterinary practices within East Anglia and North London since 1996, seeing owners of dogs, cats and rabbits.

Emma is also Features Editor of the small livestock journal *Fur & Feather*. She has a regular 'Rabbits on the Couch' column within *Rabbiting On*, the journal of the Rabbit Welfare Fund. In 2003 she helped form the Rabbit Behaviour Advisory Group (www.rabbitbehaviour.co.uk) with Georgie Hearne and Dr Anne McBride.

Georgie Hearne MSc
Georgie works as a pet behaviourist for the Animal Medical Centre, London, where she provides literature, advice and support to clients, vets and vet nurses and sees cases on veterinary referral. She writes for *Fur and Feather* magazine, and has appeared on television and radio discussing cat, dog and rabbit behaviour. Georgie has been a member of the Association of Pet Behaviour Counsellors since 2002 and is a member of the Rabbit Behaviour Advisory Group. She is also very interested in the area of human–animal interaction. She obtained her MSc in Companion Animal Behaviour from the University of Southampton in 2002.

Anthony L. Podberscek PhD
Anthony is a Postdoctoral Research Associate in the Department of Clinical Veterinary Medicine at the University of Cambridge. His veterinary degree and PhD were completed in Brisbane, Australia, and he has now been working in the UK for eleven years. He is Editor-in-Chief of the journal *Anthrozoös* (journal of human–animal interactions), a council member of the International Society for Anthrozoology, and editorial board member of the *Journal of Applied Animal Welfare Science*. Anthony is an Honorary Advisor to the APBC.

Julie Bedford BSc(Hons)
Julie has grown up and worked with dogs and horses her entire life. She has run a private behaviour practice for seventeen years ongoing and has worked for the Blue Cross for some eleven years, intitially as a centre manager, then animal behaviourist, and since 2002 as the Head of Animal Behaviour Services. Julie also lectures in

psychology for one of the largest Colleges of Further Education in the country. She has shared the last nine years with her Fell Pony, Tommy, also known as the Flying Fell due to his enthusiasm for cross-country jumping.

Donna Brander BSc(Hons)

Orginally from Texas, Donna has run behavioural practices in the USA, Europe and Hong Kong and has been settled in Edinburgh since 1994, where she was appointed as an Honorary Fellow of the Royal (Dick) Veterinary School in 1995. She has operated the consultancy from the Veterinary School since that time. In 1998 Donna was appointed Senior Lecturer in Companion Animal Behaviour, MSc course in Animal Behaviour and Welfare at the University of Edinburgh. In 1999 Donna set up The Animal Behaviour Service, Scotland with Dr Natalie Waran, Anne Gallagher-Thaw MSc, Anna Meredith BVSc MRCVS, and Samantha Scott BVSc MRCVS.

She continues to lecture on small animal behaviour both in the UK and in the USA. At the time of writing Donna is Chairman of the Association of Pet Behaviour Counsellors (APBC), a position she has previously also held. She was elected Honorary President of Dog Aid Society of Scotland in 2002.

Natalie Waran PhD

Natalie is considered to be a leading authority on horse behaviour and on how to deal with problem behaviour. She has already compiled and edited an academic book on horse welfare (Waran, N. (ed.) *The Welfare of Horses*, (2002) Kluwer Academic Publishers) as well as contributing chapters to various books (e.g., 'Social

Behaviour of Horses' in (2001), Keeling, L., and Gonyou, H. (eds) (2001) *Social behaviour in Farm Animals* CABI; and 'Training and behavioural rehabilitation in the horse', in: Mills, D., and McDonnell, S. (eds), *The behaviour of the domestic horse*, CUP (due in 2004)). She has published both in academic journals and equestrian magazines, and leads an animal behaviour and welfare research group in the Division of Veterinary Clinical Medicine at the University of Edinburgh where she runs the equine behaviour clinic.

Index